George Meghabghab and Abraham Kandel

Search Engines, Link Analysis, and User's Web Behavior

T0137754

Studies in Computational Intelligence, Volume 99

Editor-in-chief
Prof. Janusz Kacprzyk
Systems Research Institute
Polish Academy of Sciences
ul. Newelska 6
01-447 Warsaw
Poland
E-mail: kacprzyk@ibspan.waw.pl

George Meghabghab
Abraham Kandel

Search Engines, Link Analysis, and User's Web Behavior

With 114 Figures and 74 Tables

Dr. George Meghabghab
Professor of Computer Science
Roane State University
Oak Ridge, TN 37830
USA
MeghabghaGV@roanestate.edu

Dr. Abraham Kandel
Endowed Eminent Scholar
Distinguished University Research Professor
Computer Science and Engineering
University of South Florida
Tampa, FL 33620
USA
kandel@cse.usf.edu

ISBN 978-3-642-09616-7   e-ISBN 978-3-540-77469-3

Studies in Computational Intelligence ISSN 1860-949X

Cover design: Deblik, Berlin, Germany

Printed on acid-free paper

9 8 7 6 5 4 3 2 1

springer.com

Preface

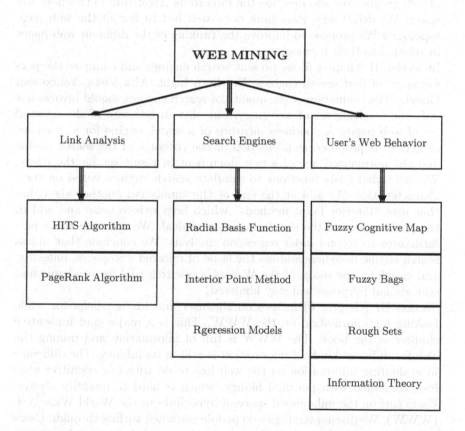

This book presents a specific and unified approach framework to three major components: Search Engines Performance, Link Analysis, and User's Web Behavior. The explosive growth and the widespread accessibility of the WWW has led to a surge of research activity in the area of information retrieval on the WWW. The three aspects of web mining follow the taxonomy of the above diagram: Link Analysis, Search engines, and User's web behavior are considered in the unifying approach. The book is organized in three sections as follows:

1. In section I of the book (chapters 2–4) we study Link Analysis within the hubs and authorities framework. Link Analysis is the science of hyperlink structures ranking, which are used to determine the relative authority of a Web page and produce improved algorithms for the ranking of Web search results. We use the HITS Algorithm developed by Kleinberg and we propose to study HITS in a 2-D new space: In-degree and Out Degree variables. After we categorize each web page into a specific toplogy we study the impact of each web topology on HITS in the new 2-D space. We describe why HITS does not fare well in almost all the different topologies of web graphs. We also describe the PageRank Algorithm in this new 2-D space. We detail why PageRank does well but in few of the web page topologies We propose to improve the ranking of the different web pages in relation to their topologies.

2. In section II (chapter 5) we present search engines and compare the performance of four search engines: Northern Light, Alta Vista, Yahoo and Google. The evaluation of the quality of search engines should involve not only the resulting set of web pages but also an estimate of the rejected set of web pages. A goodness measure of a search engine for a given set of queries is proposed as a function of the coverage of the search engine and the normalized age of a new document in result set for the query. We use radial basis functions to simulate search engines based on their characteristics. We add at the end of the simulation another algorithm that uses "Interior Point methods" which help reduce noise and add to the performance of the simulation a great deal. We compare those performances to second order regression analysis. We conclude that unless search engine designers address the issue of rejected web pages, indexing, and crawling, the usage of the Web as a research tool for academic and educational purposes will stay hindered.

3. Section III (chapter 6) focuses on the user and his/her behavior while looking for information on the WWW. This is a major and innovative chapter of the book. The WWW is full of information and mining the web for different kinds of information is still in its infancy. The difficulty in evaluating information on the web has to do with our cognitive abilities combined with personal likings, which is hard to quantify always. Users surf on the unbounded space of hyperlinks or the World Wide Web (WWW). We discuss strategies do people use when surfing through. Users form the cognitive neurological networks that result in a mental pathway, or cognitive map, that makes more navigable the route to further information as well as the information they set out to find. We propose three new innovative approaches to quantify user's behavior:

(a) Fuzzy cognitive maps (FCM) can model the causal graph of "expert's perception" of how users behave on the web. We establish a stable dynamical system that behaves well in a limit cycle. When the original FCM causal system is left to behave in time, the system learn new rules or reinforce old ones. If Differential Hebbian Learning is used (DHL), the system uncovers new rules in the case of failure and in the case of success. That shows that the original causal matrix can uncover some dynamical rules on learning when left to behave in time, which was not apparent in the beginning. The dynamical system can move from failure to success with certain conditions imposed on the user. The same rules that switch users from failure to success can keep successful users successful.

(b) In all studies of user's behavior on searching the web, researchers found that their search strategy or traversal process showed a pattern of moving back and forth between searching and browsing. These activities of searching and browsing are inclusive activities to complete a goal-oriented task to find the answer for the query. The moving back and forth, including backtracking to a former web page, constitutes the measurable or physical states or actions that a user takes while searching the web. The structure of a bag as a framework can help study the behavior of users and uncover some intrinsic properties about users. We consider queries that depend on one variable or "uni-variable". An example of "uni-variable" queries is: "Find the number of web pages of users whose searching is less than the average". Fuzzy operators were used successfully in comparing the "fuzzy bag" of the successful users to the "fuzzy bag" of unsuccessful users to answer such a query. The research considered queries that depend on more than one variable or multi-variable. An example of a "multi-variable" query is: "Find the number of web pages of users whose number of searches is less than average and hyperlink navigation is less than the average". One-dimensional fuzzy bags were extended to two dimensional fuzzy bags and two dimensional fuzzy bag operators were used to answer such a query.

(c) We turn the comparison of successful students to unsuccessful students on a specific factual query into a decision table of twenty rows. Rough Sets offer a unique framework to extract rules from the decision table. We consider a decision table with three set of attribute variables:

 i. The case of the 3 condition variables or attribute variables: Searches(SE), Hyperlinks(H), Web Pages(W). We reduce the number of attributes and generate the rules that explain the behavior of the users using these 3 variables. We discover that these two variables H and W are just sufficient to make the decision as whether a user succeeded or failed. The attribute SE is become superfluous. Six deterministic rules with an average of 1.67 conditions per rule were extracted and two non deterministic rules.

    ii. The case of the 3 condition variables or attribute variables: Web Pages(W), Time(T), Searches(SE). We discover that these three variables were needed to make a decision either a user succeeded or failed. Seven deterministic rules with an average of 1.55 conditions per rule were extracted and two non deterministic rules.

   iii. The case of the 4 condition variables or attribute variables: Web Pages(W), Hyperlinks(H), Time(T), Searches(SE). We discover that these four attributes could not be reduced without comprising making a decision whether a user is successful or failed. Nine deterministic rules with an average of 1.77 conditions per rule were extracted and one non deterministic rule.

(d) We turn the comparison of successful students to unsuccessful students into a decision table of twenty rows but we used this time Information Theoretic (ID$_3$) framework. We compare the results of Rough Sets to ID$_3$ in the above three cases 3(c)i through 3(c)iii. Rough sets performed better than ID$_3$ in all three cases 3(c)i through 3(c)iii.

Four mathematical complements expand on needed theoretical background and present descriptions of very specific examples including detailed results. The book provides both theoretical and practical coverage of all three major components introduced above:

1. Includes extensive number of integrated examples and figures.
2. Include an end of chapter summary, and plenty of examples to cover the ideas discussed.
3. Assumes a modest knowledge in linear algebra, fuzzy sets, fuzzy bags and statistical analysis.
4. Fits nicely in an upper level undergraduate course on the Web Mining.

This book can be used by researchers in the fields of information sciences, engineering (especially software), computer science, statistics and management, who are looking for a unified theoretical approach to finding relevant information on the WWW and a way of interpreting it from a data perspective to a user perspective. It specifically stresses the importance of the involvement of the user looking for information to the relevance of information sought to the performance of the medium used to find information on the WWW. In addition, social sciences, psychology, and other fields that are interested in understanding the underlying phenomena from data can benefit from the proposed approach. The book can also serve as a reference book for advanced undergraduate/graduate level course in data mining on the WWW. Practitioners may be particularly interested in the examples used throughout the book.

USA   *George Meghabghab and Abraham Kandel*
April 2007

Contents

List of Figures

List of Tables

1

Basic WWW Technologies

1.1 General Introduction

Internet search service google.com announced recently that its indexes span an unprecedented 8 billion Web documents. The total includes a few billion Web pages, few hundreds of millions of images and about a decade's worth - some 1 billion - of newsgroup messages. Promoting technology that scours this vast collection in less than a second, Google says that obtaining a similar result by hand would take a human searcher 15180 years - assuming someone could be found to scrutinize one document every minute, for twenty-four hours per day. If we were to put 15180 humans to perform this task we could have done it in one year. To do it in a minute we have to use 5.6 million people and to do it in a second like google does we have to use 333 million people. That is just 1 in every 20 people on the face of the earth will be just scrutinizing the 8 billion Web pages that google can index for now. As stark as this illustration may be, it points to the burgeoning scale of Web-based information resources. In a few short years, the Web has become our most compelling technological accomplishment. It's the world's largest library and telephone network, the world's largest jukebox and photo album. It is global, and at the same time perfectly local. It's the fastest growing enterprise on Earth, and yet, no stock-broker or media pundit or taxman can tell us who's actually in charge. Social scientist *Ithiel de Sola Pool* [27], in a posthumous 1990 collection "Technologies Without Boundaries," called the Internet "part of the largest machine that man has ever constructed." A dozen years later, de Sola Pool appears eloquent in his understatement. As we learn to use this machine for more than merely searching an impending googolplex of indexed terms, our ability to repurpose, aggregate, and cross-reference individual datum inexpensively offers new impetus to economic and societal velocity. Where today the Web presents a loosely-coupled periodical file or a waste-bin of one-way screeds, we are discovering ways to unearth semantic clues that yield new knowledge, fresh applications and, perhaps, astounding insights. For enterprises doing business on the Web, the network is also becoming the world's largest cash

G. Meghabghab and A. Kandel: *Search Engines, Link Analysis, and User's Web Behavior*, Studies in Computational Intelligence (SCI) **99**, 1–22 (2008)
www.springerlink.com   © Springer-Verlag Berlin Heidelberg 2008

register, a development that will only accelerate as we find new ways to mine value from the content institutions and individuals place there. Information architects use the term Web mining to describe the process of locating and extracting usable information from Web-based repositories and doing so on a repeatable, continuous basis. In this book, we describe this new discipline and explain "hubs" and "authorities" on the Web, google's Web ranking to arrive at better mining the Web for information as well as users that understand better their needs.

1.1.1 The Internet is Big

In a decade following release of the NCSA Mosaic Web browser, the World-wide Web has become the world's de facto information repository. For 2002, internet marketing statistician eMarketer estimates that there are already 168 million internet users in North America, lagging slightly the 175 million users in Europe and easily doubling Asia's estimated 85 million users. With Asia pacing the 38% growth in these numbers, we can expect some 590 million Web-surfing individuals in the three regions by 2004. Despite public perceptions that the internet juggernaut is fueled by hype, the scope and scale of this deployment is unprecedented in human experience. Today, virtually no one disputes that the worldwide Web is becoming an integral component in the way advanced societies give and get information. Here are just a few measures of the Web's utility to consumers:

1. 85% of North American internet users report having bought goods or services online within the past year (up from 77% a year ago)
2. Online retail sales in the US totaled $51.3 billion in 2001 and are expected to jump 41% to $72.1 billion this year.
3. 49% of North American internet users have conducted a bank transaction online (up from 29% a year ago)
4. 19% of all US people aged 12 and older report having downloaded digital music files from an online file-sharing service (that's 41.5 million people)
5. 13% have invested online (up from 10% a year ago).

These numbers are dwarfed, however, by internet-enabled activity by and between businesses. Whether explicitly global or durably local, growth in business-to-business ecommerce is exploding. eMarketer projects that world-wide business-to-business (B2B) ecommerce revenues will grow by almost 74 percent in 2002. From a base of $278 billion in 2000, B2B revenues will reach $823.4 billion by the end of this year and eclipse the trillion-dollar mark in 2003. All this activity, of course, generates new data. A recent University of California-Berkeley study that found that the world's annual production of information exceeds 250 megabytes per living person, the largest proportion of which now first appears in digital form, making it a candidate for inclusion in Web-based applications. Whether we're talking about transaction details along an industrial supply chain, bid/asked quotations in an electronic market, or customer complaints through a CRM portal, much of this content either originates in or is destined for a structured database management system and its dependent applications. There - safely defined in rows and columns and validated against application and data-type logic - it can be queried, extracted, extrapolated, or conjoined in ways that the original program designers scarcely contemplated. Such is the power of structured data management tools. What is less easily accomplished is sharing this data beyond an enterprise's firewalls again in anything other than static form. Every commercial database reporting tool now generates Web browser output, but it stands as one of the great ironies of the relational database era how so much of any system's output becomes unstructured again at the first point where it actually becomes useful to someone. Unlike the orderly universe of database management, Web mining is concerned with the detection, capture and subsequent re-use of information content that we first encounter "in the wild," that is to say, in a Web-page presentation accessible via browser. Thanks to their ubiquity and low cost, browsers have become our indispensable keyholes into enterprise- and customer-facing applications, with the result that a growing proportion of business data are destined to pass through one at some point during the digital half-life. A primary premise of Web miners is that this content is "low-hanging fruit" that can be harvested and re-used. Where do we find the low-hanging fruit? We use Web mining when:

1. We need access to external counter-party or partner's data.
2. When we need to aggregate many values from sources that are spread across independent systems.
3. When our applications require loosely-structured or unstructured data, such as text in headlines or resumes.

1.1.2 Playing Smart in a Global Trade

Global energy markets offer significant pricing and strategic leverage to participants who come to trade armed with well-head production figures or tanker routing or other operational data that support accurate commodity value discovery. Specialized (and profitable) information services have long existed to satisfy this demand, which they accomplish by gleaning breaking information from market participants, then filter it through analytical models and broadcast it over subscriber newsfeeds. Today, energy producers routinely record notices of vessel sailing dates or production data on their own Websites, making the collection and assembly of up-to-date information a matter of careful diligence more than hardcore sleuthing. This process is complicated, however, by the sheer number of players involved and the frequent changes in critical information. Web mining provides these specialized publishers a way to scale their market coverage and improve the timeliness and accuracy of services without adding new researchers.

1.1.3 EDI for the Rest of Us

"Just-in-time" manufacturers in the semiconductor and electronics industries publish catalog or inventory data on their Websites as a pragmatic way to accomplish non-invasive EDI inexpensively. They've learned that monitoring "loosely-coupled" supply chain relationships through a partner's finished-goods or parts inventories facilitates their own ordering, production planning or billing processes. Unfortunately, most users of this data find themselves manually extracting information (and more often that not, re-keying it) from their partners' disparate Websites. One businessman described the dilemma this way: "We have access to our most-critical suppliers' inventories and use this information constantly in our production planning and purchasing decisions. Problem is, our divisions, product-line managers and departmental functions all have different requirements and there isn't any single business application that consolidates a global view of our marketplace in a way that suits everybody. If we had just one supplier, we'd "hardwire" our systems to his and get everything we need in real-time. But we've identified 40 key external participants in our production processes and putting together a "global" view of our supply-side marketplace from, say, a Financial or Purchasing or Production Control or TQM perspective is still largely a matter of cut-and-paste. It's a lot of handwork." Automated Web mining tools allow much closer, round-the-clock scrutiny of vertical supply chains and eliminate the handwork.

Partners can assemble a complete view of the components used in manufacturing processes and make better decisions about re-supply agreements, accounts payable issues or potential sources of production delays. Where "short-run" production relationships or nimble marketing strategies make hard-wiring EDI protocols difficult, Web mining can mean the difference between operating efficiently and operating in the dark.

1.1.4 Merger-Proofing Your Business Intelligence

Global enterprises doing business in distant markets have long understood the imperative to know their customers thoroughly. That means anticipating customer demand, correlating services across a history of customer interactions, and staying atop developments in a customer's downstream markets. CRM software provides a powerful means to track all this data, but it requires interaction with separate accounting and order-management systems, customer call centers, and business intelligence applications both within and outside an organization's firewalls. Where all the pieces exist in a single enterprise, this is a significant integration chore. But add the urgency of mergers or other strategic business reorganizations and this hard-won transparency is often the first element to fail. Companies that know everything about their markets are the first to discover how little they know about customers acquired in a business combination. Web mining is a pragmatic way to fill in those blanks, incorporating data that would otherwise require complicated software integration and expensive marketplace assessments by an army of consultants. Whether viewed as a pragmatic solution to information architecture problems or as a new approach to automating business intelligence, Web mining unlocks valuable nuggets of information from the looming info-glut and delivers essential content to the heart of critical business processes.

1.1.5 Technical Challenges Abound

General-purpose Web mining applications are, of course, only just now becoming feasible. Where Web applications have made great strides in improving the timeliness or accessibility of data across the extended enterprise, browsers and visual representation remain their primary output state. Getting specific content out of the browser and into an application's database, or a spreadsheet model, or an email alert, or a mobile employee's wireless device is the new challenge. Where data must be exchanged, re-purposed for new uses, that visual representation we experience in our browser sessions becomes a virtual speed-bump. The manual cut-and-paste experience shared by nearly everyone who has ever circulated an internet joke or captured a URL in a browser bookmark offers a conceptual foretaste of Web mining. Simple as it is to understand though, Web mining is very difficult to replicate in automated processes. That's because the Web is not constructed like a database, where standard access- and data-type rules, sharing conventions or navigational procedures enforce predictability. As implemented, the Web is the Wild, Wild West, ruled not by gunslingers, but by graphical considerations and constant ad hoc construction that long ago defied machine parsing. In "Digerati: Encounters With the Cyber Elite" by *John Brockman* [14], former Thinking Machines Corporation founder Danny Hillis writes: "When computers pass around Web pages, they don't know what they're talking about. They're just passing around bits, and they can be meaningless bits as far as they're concerned. They're just acting like a big telephone system. It's much more interesting when computers actually understand what they're talking about." A fierce observer of modern media and the societies they serve, *Marshall McLuhan* [79] would have loved the worldwide Web. As a result of the Web's early dominance, the internet today exists largely as a simple global conduit connecting human audiences to rich graphical or multimedia content organized around human sensory expectations. Web designers have seen their craft evolve from composing static pages that must be created, published and stored ahead of any authentic user-demand, to dynamic applications that interoperate with backend databases, instrumentation packages and contextual information supplied by viewers. State-of-the-art Websites composite many applications or data sources into aggregated "portal" pages and populate them freshly for us, on the fly. These applications are richer by far, but they are fundamentally tied to our human capacity to inquire, to comprehend, and to absorb. In part, that's because the internet as we know it is designed to be keyboard-driven, transmuting mouse clicks, button and scrollbar functions and visible hypertext clues into the logic that defines our interaction with a distant server. Even where content is audible (as with streaming media) or report-structured (as results from the tabular output of database inquiries), we still navigate by pages, borrowing a visual publishing metaphor older than moveable type. All information on the Web thus takes its shape in conformance with the tyranny of browsers. Since this is exclusively a presentation-layer service, the horsepower trapped in millions of desktop computers does little more than host session-level page presentations, squandering one of the largest peacetime capital investments in the history of civilization. Unfortunately, until applications routinely comply with industry-wide peer-to-peer computing standards, machines will do a poor job parsing Web-based applications for intelligence. That's because today's graphically-rich Web applications make use of HTML extensively, commingling actual data content with display and formatting instructions that are meaningful only when a page is rendered for human consumption. A generation of graphics designers has taken to heart the lessons of *Edward Tufte's* [128] "The Visual Display of Quantitative Information," turning tables of dull numbers into rich visual metaphors, all to better stimulate our cognitive appreciation of the relationships they reveal. We know what thy mean when we see them and, driven by concern for our cognitive bandwidth, Web designers take generous liberties composing pages that "look right." They routinely pad tables with non-displaying characters or deviate from previous style sheet declarations to make objects "fit" visually. If this were as bad as it gets, the machine-based parsing tools that make Web mining practical could be assumed to eventually conquer the problem with brute computational force. But it's much worse than that. Brute force presumes the existence of standard text labels from standard glossaries, or at worst precise positional repetition throughout a file. What is a machine to do, for instance, when a five-digit numeric sequence that one Web-designer labels "mail:" is pre-pended on another site by a graphical object or? Are they both zip codes? If simplicity and broad application across hundreds or thousands of Websites is our goal, the need to record complex parsing rules for every such example defeats our purpose. Positional data, too, is meaningless when Web pages routinely adjust themselves to accommodate browser versions or even a user's screen settings. It's fair to say that business transaction records - arguably among some of the most valuable content on the Web - change too frequently for such simple screen-scraping techniques to work reliably. The problem becomes even more difficult when information is dispersed in bits and pieces all over a Website. Web mining tools need to be able to follow links and perform multi-page aggregations. Secure access provisions also pose issues for software that must scour a distant site at regular intervals. Getting through password authentication or firewalls is intentionally difficult. As if these challenges weren't enough, much of the information destined for Web mining applications is stored and displayed in other formats, such as PDF files. A robust Web mining software solution is one that recognizes content within such "deep Web" structures and, similar to parsing HTML, converts markup characteristics into semantic structure.

1.2 Web Documents

"The Web has grown into a vital channel of communication and an important vehicle for information dissemination and retrieval." (Martzoukou [77]). Whilst more than one would be persuaded that information retrieval on the Web is easy and at hand with a simple click of the mouse, still, the situation is not as bright. Information retrieval on the Web is far more difficult and different from other forms of information retrieval than one would believe since Web documents (pages) are voluminous, heterogeneous and dynamic with a lot of hyperlinks including useless information such as spams, have more than a dozen search engines, make use of free text searching, are mostly without metadata and standards, and contain multimedia elements. The principles of searching a catalogue are different from that of a WWW and the WWW is devoid of cataloguing and classification as opposed to a library system. Even the evaluation of information retrieval on the Web is different. All these points contribute to make information retrieval on the Web different from say, a library catalogue or even an electronic database. "Information retrieval deals with the representation, storage, organization of and access to information items. The representation and organization of the information should provide the user with easy access to the information in which he/she is interested ([4]). Information retrieval on the Web differs greatly from other forms of information retrieval and this because of the very nature of the Web itself and its documents. Indeed, "the WWW has created a revolution in the accessibility of information" (National Information Standard Organization, 2004). It goes without saying that the Internet is growing very rapidly and it will be difficult to search for the required information in this gigantic digital library.

1.2.1 Information Overload

Information overload can be defined as "the inability to extract needed knowledge from an immense quantity of information for one of many reasons" (Nelson [105]). One of the main points that would differentiate information retrieval on the Web from other forms of information retrieval is information overload since the Web would represent items throughout the universe. "The volume of information on the Internet creates more problems than just trying to search an immense collection of data for a small and specific set of knowledge" ([105]), An OPAC will only represent the items of a specific library or a union catalogue will only represent items of participating libraries within a region or country. It is known that large volumes of data, especially uncontrolled data, are full of errors and inconsistencies. Contrast the situation on the Web by *Nelson* [[105]], "when we try to retrieve or search for information, we often get conflicting information or information which we do not want"; and by *Meghabghab* [90] "finding authoritative information on the Web is a challenging problem" to the situation of a library where we would get both authoritative and relevant information. Information overload on the Web makes retrieval become a more laborious task for the user. The latter will either have to do several searches or refine searches before arriving at the relevant and accurate document. Therefore, as opposed to a search in a catalogue, with the Web, there is no instantaneous response as such. There is no instantaneous response in the sense that there are no instant relevant results and this is not to be mixed up with instantaneous response in the sense that it is true that whilst we tap in keywords, we do get a thousand of hits. With the advent of the Web, users will need easier access to the thousands of resources that are available but yet hard to find.

1.2.2 Hyperlinks

"With the advent of the Web, new sources of information became available, one of them being hyperlinks between documents" (Henzinger et al. [53]). In an OPAC for example, one would not find hyperlinks. The Web would be the only place to find hyperlinks and this brings a new dimension to information retrieval. With hyperlinks, one information leads to another. If the user were viewing a library catalogue, he/she would be directed to let's say a book and that's the end to it. But if that same user does a search on the Web, he/she would be presented with a thousand of hits, he/she would then need to choose what to view and then the Web pages viewed would refer the user to yet other pages and so on. As pointed out in [53], "hyperlinks provide a valuable source of information for Web information retrieval" hyperlinks are thus providing us with more or even newer information retrieval capabilities.

1.2.3 Heterogeneous Nature of the Web

"A Web page typically contains various types of materials that are not related to the topic of the Web page" (Yu et al. [133]). As such, the heterogeneous nature of the Web affects information retrieval. Most of the Web pages would consist of multiple topics and parts such as pictures, animations, logos, advertisements and other such links. "Although traditional documents also often have multiple topics, they are less diverse so that the impact on retrieval performance is smaller" (Yu et al. [133]). For instance, whilst searching in an OPAC, one won't find any animations or pictures interfering with the search. Documents on the Web are presented in a variety of formats as opposed to catalogues and databases. We have HTML, pdf, MP3, text formats, etc. The latter can be barriers to information retrieval on the Web. For one to retrieve a pdf document on the Web, one must have Acrobat Reader software installed and enough space on the computer's hard disk to install it.

1.3 Search Engines

A search engine is a very complex machine. To engineer one is a very challenging task. From the early days of yahoo, and google, creating a search engine which scales to then's Web presented many challenges. Fast crawling technology is needed to gather the Web documents and keep them up to date. Storage space must be used efficiently to store indices and, optionally, the documents themselves. The indexing system must process hundreds of gigabytes of data efficiently. Queries must be handled quickly, at a rate of hundreds to thousands per second. As opposed to other forms of retrieval, the Web makes use of sophisticated information retrieval tools known as search engines. A search engine is simply "a Web site used to easily locate Internet resources". Search engines have facilitated the information retrieval process by adopting techniques such as Artificial Intelligence, Bayesian Statistics and probability theory, weighting and also, query by example (see Mathematical Complements) . We can add that without search engines, information retrieval would be impossible. The Web relies upon search engines just like libraries rely upon catalogues. These tasks are becoming increasingly difficult as the Web grows. However, hardware performance and cost have improved dramatically to partially offset the difficulty. There are, however, several notable exceptions to this progress such as disk seek time and operating system robustness. A search engine would consist of three parts, namely, an interface, an index and the Web crawler or spider. The interface is the Web page where one would normally formulate his/her queries whereas the index would be the database operating behind the Web page. The crawlers or spiders are programs that would crawl throughout the Web, visiting each site and gathering information. The role of the search engine is to provide more control for the user in performing a search. Those search engines make use of the index to fetch terms of the query. The higher the data in the index, the higher would be the number of hits. However, the size of the index would vary from search engine to search engine but the bigger the index the better and the more often it is updated the better. Search engines are different in nature to electronic databases or library catalogues. Search engines would include a number of free Web pages from around the world since no search engine would include every Web page whereas electronic databases would include citations to some of the articles published in a particular subject or journal. The latter may be a fee-based service. Library catalogues of whatever format would record the items of libraries. The advantage with electronic databases and library catalogues is that they are regularly updated whereas search engines do not have a definite timeframe as to when they are updated and contain links that no longer exist. Statements as those of *Hersh* [55], "unlike MEDLINE and CD-ROM textbooks, the Web is not a single database". Therefore, given the size of the Web, no single search system will be able to search the entire Web. As far as the searching process is concerned, search engines would make use of keywords or subject headings that they have themselves established whereas library catalogues would make use of standard subject headings such as the Library of Congress Subject Headings. Carrying out a search using a search engine would be different from that of a catalogue in the sense that we would hardly find any fields for titles, names and subjects in a search engine. We can do a title search using a search engine but it will operate on the title as is the title and not all Web sources have proper titles. Whilst searching in a library catalogue, knowing a title is enough and we can get to the source directly. But nevertheless, with a search engine, the depth of searching is deeper than with a library catalogue. For instance, search engines would search parts of a book if the book is available on the Web but a library catalogue would search the book as a whole and it is up to the user to find the relevant chapters. Search engines also present the hits in order of relevance but it is to be noted that only the person doing the search can judge the relevance of a document. While moat OPACs would present result sets sorted by either author, title or in chronological order. As opposed to other forms of information retrieval tools, search engines provide us with more up to date information at the click of a mouse despite the fact that not all the information are useful ones. Once someone writes a piece of a material and puts it online, anyone in the world will be able to reach it. In designing google for example, *Brin and Page* [12] have considered both the rate of growth of the Web and technological changes. From the early days, google was designed with the intentions of scaling well to extremely large data sets. It makes efficient use of storage space to store the index. Its data structures are optimized for fast and efficient access (see figure 1.1). Further, they expected that the cost to index and store text or HTML will eventually decline relative to the amount that will be available. This will result in favorable scaling properties for centralized systems like google. In google, the Web crawling (downloading of Web pages) is done by several distributed crawlers. There is an URLserver that sends lists of URLs to be fetched to the crawlers. The Web pages that are fetched are then sent to the storeserver. The storeserver then compresses and stores the Web pages into a repository. Every Web page has an associated ID number called a docID which is assigned whenever a new URL is parsed out of a Web page. The indexing function is performed by the indexer and the sorter. The indexer performs a number of functions. It reads the repository, uncompresses the documents, and parses them. Each document is converted into a set of word occurrences called hits. The hits record the word, position in document, an approximation of font size, and capitalization. The indexer distributes these hits into a set of "barrels", creating a partially sorted forward index. The indexer performs another important function. It parses out all the links in every Web page and stores important information about them in an anchors file. This file contains enough information to determine where each link points from and to, and the text of the link. The URLresolver reads the anchors file and converts relative URLs into absolute URLs and in turn into docIDs. It puts the anchor text into the forward index, associated with the docID that the anchor points to. It also generates a database of links which are pairs of docIDs. The links database is used to compute PageRanks for all the documents. The sorter takes the barrels, which are sorted by docID resorts them by wordID to generate the inverted index. This is done in place so that little temporary space is needed for this operation. The sorter also produces a list of wordIDs and offsets into

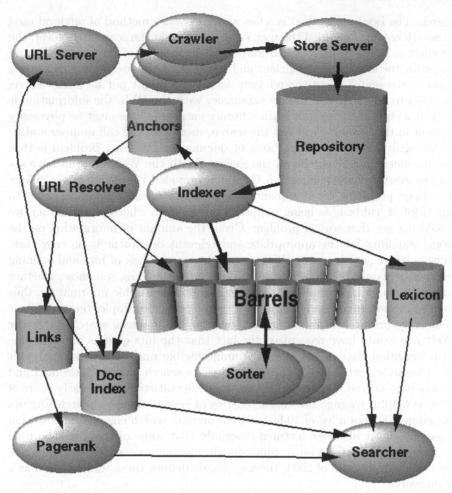

Fig. 1.1. High Level google Architecture

the inverted index. A program called DumpLexicon takes this list together
with the lexicon produced by the indexer and generates a new lexicon to be
used by the searcher. The searcher is run by a Web server and uses the lexicon
built by DumpLexicon together with the inverted index and the PageRanks
to answer queries. The principles of searching a library catalogue, is also very
different from carrying out a search on the Web. We are familiar with the
library catalogue which would consist of a call number, a title or author or
subject entry and all these in a standardized format. If the user knows one of
this information, then the user will be able to retrieve the exact information.
However, searching the Web would depend upon keywords and the Boolean
operators and, "OR" and "NOT" in order to either broaden or narrow a

search. The Boolean method is a fast and fairly easy method of retrieval used in search engines provided the user knows how to do the search. However, the problem is that the user should have some knowledge of the search topic in order for the search to be efficient and effective. If the user enters the wrong term in the search, then the relevant documents might not be retrieved. As opposed to a search in a library catalogue, with the Web, the information is only at a click of the mouse. With a library catalogue, one must be physically present in the library, carry out the search, memorize the call number and go to the shelves to retrieve the book or document. The next problem is that the document might not be on the shelves. With the Web, one can have access to every other document on the globe provided the document is online. The main problems are the quantity and quality of information. That is to say, a lot of 'rubbish' is being published on the Web whereas in a library, we would not get that sort of problem. Given the amount of information on the Web, searching for the appropriate and relevant document is no easy task. With a single search on the Web, one might get millions of hits and opening every page would be time-consuming. Library catalogues, somehow, whether online or manual, provide the user with more bibliographic information, thus facilitating information retrieval. In a catalogue for example, one would get some very brief bibliographic information for the material searched. On the Web, one would have recognized the fact that the hits give us only one line of information that is at times even unintelligible until you click on the hit in order to see what it is. Through the years search engines have died and others have stayed the course. Yet some of the statistics of the early years of 2000 is helpful to gage the characteristics of search engines. Search Engines showdown [1] kept a list of 10 features to compare search engines. This section looks closely at all these features especially that some of them will be used in more details in Chapter 5. Since google has dominated the search engine showdown in the mid of 2004, the way google defines these terms is used as a framework.

1.3.1 Listed Features

1. Defaults: What happens when multiple terms are entered for a search using no Boolean operators, + or − symbols, phrase marks, or other specialized features. Example: two terms could be processed as two AND terms two OR terms or "two terms" as a phrase. In google for example, multiple search terms are processed as an AND operation by default. Phrase matches are ranked higher
2. Boolean: refers to how multiple terms are combined in a search. "AND" requires that both terms be found. "OR" lets either term be found. "NOT" means any record containing the second term will be excluded. ( ) means the Boolean operators can be nested using parentheses. + is equivalent to AND, requiring the term; the + should be placed directly in front of the search term. - is equivalent to NOT and means to exclude the term;

the - should be placed directly in front of the search term. Operators can be entered in the case shown by the example. Example: "(Meghabghab and (authorities or hubs) and not Kleinberg)" returned 34 hits in google (March 28, 2007). while "(Meghabghab and (authorities or hubs))" returned 247 hits (March 28, 2007). Google always searches for pages containing all the words in your query, so you do not need to use + in front of words.

3. Proximity: refers to the ability to specify how close within a record multiple terms should be to each other. The most commonly used proximity search option in Internet finding aids is a "phrase search" that requires terms to be in the exact order specified within the phrase markings. The default standard for identifying phrases is to use double quotes (" ") to surround the phrase. Phrase searching example: "soul searching is good" in google returned 35 hits. Beyond phrase searching, other proximity operators can specify how close terms should be to each other. Some will also specify the order of the search terms. Each search engine can define them differently and use various operator names such as NEAR, ADJ, W, or AFTER. Example: Meghabghab NEAR Kleinberg returned 23 hits in google (March 28, 2007).

4. Truncation and Stemming: Truncation refers to the ability to search just a portion of a word. Typically, a symbol such as the asterisk is used to represent the rest of the term. End truncation is where several letters at the beginning of a word are specified but the ending can vary. With internal truncation, a symbol can represent one or more characters within a word. While stemming related to truncation, usually refers to the ability of a search engine to find word variants such as plurals, singular forms, past tense, present tense, etc. Some stemming only covers plural and singular forms. End Truncation Examples: Meghabgha* finds 3 hits in yahoo for Meghabghab (nothing else) but none in google since the * feature did not work in google (April 6, 2005). Since then google added that feature and as of June 6, 2006 Meghabgha* returned 3 hits in google. Truncation Examples: Megha*ghab finds none in yahoo. In google, as of June 6, 2006, one hit is returned. Yet, if you were to type "soul * is good" yahoo returned 9600 hits among with the following phrase ranking second "Like cold water to a weary soul, so is good news from a far country", while google returned 61,500 hits with "As cold waters to a faint soul, so is good news from a far countrysoul searching is good" ranked fourth.

5. Case: In general, most search engines will match upper case, lower case, and mixed case as all the same term. Some search engines have the capability to match exact case. Entering a search term in lower case will usually find all cases. In a case sensitive search engine, entering any upper case letter in a search term will invoke the exact case match. Example: meghabghab, Meghabghab returns equal hits while MEGHABGHAB returns more hits in yahoo. Google has no case sensitive searching.

6. Fields: Fields searching allows the searcher to designate where a specific search term will appear. Rather than searching for words anywhere on a Web page, fields define specific structural units of a document. The title, the URL, an image tag, and a hypertext link are common fields on a Web page. Example: title:Meghabghab will look for the word 'Meghabghab' in the title of a Web page. yahoo returns 30 hits. While google uses intitle to get similar results to that of yahoo. Example: intitle:Meghabghab returns 60 Hits in google (June 6, 2006).

7. Limits: The ability to narrow search results by adding a specific restriction to the search. Commonly available limits are the date limit and the language limit. The latter would restrict the search results to only those Web pages identified as being in the specified language. google has language, domain, date, filetype, and adult content limits. The date limit, added in July 2001, is only available on the Advanced Search page. Only three options are available: Past 3 Months, Past 6 Months, or Past Year. Meghabghab past 3 months returns 11,400 hits (as of March 28, 2007), returns 11,300 hits for past 6 months (as of March 28, 2007), and returns 11,400 hits for the past year (as of March 28, 2007). As for the file type you can specify a file type: Meghabghab filetype:pdf returns 164 hits in google.

8. Stop Words: Frequently occurring words that are not searchable. Some search engines include common words such as 'the' while others ignore such words in a query. Stop words may include numbers and frequent HTML strings. Some search engines only search stop words if they are part of a phrase search or the only search terms in the query. Examples: the, a, is, of, be, l, html, com.

9. Sorting: The ability to organize the results of a search. Typically, Internet search engines sort the results by "relevance" determined by their proprietary relevance ranking algorithms. Other options are to arrange the results by date, alphabetically by title, or by root URL or host name. In google, results are sorted by relevance which is determined by google's PageRank analysis, determined by links from other pages with a greater weight given to authoritative sites. Pages are also clustered by site. Only two pages per site will be displayed, with the second indented. Others are available via the [More results from · · ·] link. If the search finds less than 1,000 results when clustered with two pages per site and if you page forward to the last page, after the last record the following message will appear: In order to show you the most relevant results, we have omitted some entries very similar to the 63 already displayed. If you like, you can repeat the search with the omitted results included. Clicking the "repeat the search" option will bring up more pages, some of which are near or exact duplicates of pages already found while others are pages that were clustered under a site listing. However, clicking on that link will not necessarily retrieve all results that have been clustered under a site. To see all results available on google, you need to check under each site cluster as well as using the "repeat this search" option.

10. Family Filters: The ability to limit or filter search results to exclude adult or other material that may be objectionable. In particular, these are made available to exclude results deemed potentially inappropriate for viewing by children.

1.3.2 Search Engine Statistics ([1])

1. Size: The size feature compared nine search engines, with MSN Search and HotBot representing the Inktomi database. This analysis used 25 small single word queries. google found more total hits than any other search engine. In addition, it placed first on 25 of the 25 searches, more than any of the others and the first time that any search engine placed first on every single search. AlltheWeb moved back into second place with significant growth since March. AltaVista also had significant growth and moved up to third. WiseNut dropped to fourth and HotBot is up to fifth. Despite sharing an Inktomi source, HotBot found more than MSN and included PDF files not available from MSN. Note that the number given in the chart reflects only the number of Web pages found, not the total number of results. The chart in figure 1.2 gives the total verified number of search results (including Web pages, PDFs, other file types, and even google's unindexed URLs) from all 25 searches. Since the exact same queries were used in March 2002 and August 2001, the other columns give previous totals.

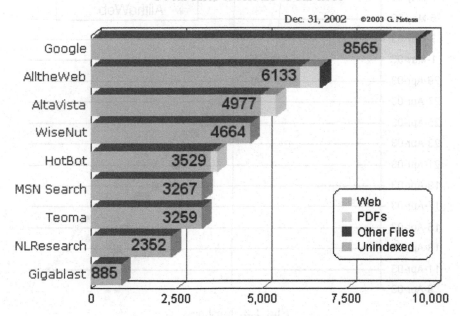

Fig. 1.2. Total Hits from 25 searches

2. Freshness: All search engines are pictures of the past, but which search
   engine has taken its picture most recently? This comparison tries to begin
   to answer that question and tracks how the search engines change over
   time. Key points from results:

   (a) Most have some results indexed in the last few days
   (b) But the bulk of most of the databases is about 1 month old
   (c) And some pages may not have been re-indexed for much longer

   This freshness showdown evaluated the search engines based on 6 searches
   (with 10-46 total useful matches per search engine). The useful matches
   were hits for specific pages that:

   (a) Are updated daily
   (b) And report that date

   And the reported date is also visible from the search engine results. The
   chart in figure 1.3 shows how old the most recently refreshed pages were

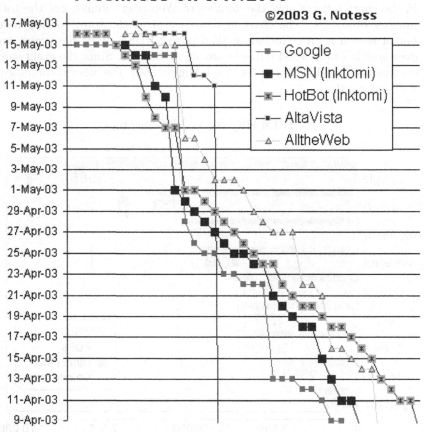

Fig. 1.3. Freshness

and the age of the most ancient page retrieved. All URLs analyzed showed the current date at the time of the test, but the search engines had all indexed older versions. Since the freshest search engines are so crowded at the top, here is another graph with a close-up of the five most current databases looking at the recent hits from about a month and a half.

3. Overlap: This analysis compares the results of four small searches run on ten different search engines. The four searches found a total of 334 hits, 141 of which represented specific Web pages. Of those 141 hits, 71 were found by only one of the ten search engines while another 30 were found by only two. Several of the largest search engines have shown significant growth since Feb. 2000, when the overlap comparison was last run. Even so, almost half of all pages found were only found by one of the search engines, and not always the same one. Over 78% were found by three search engines at most. Each pie slice in the chart (see figure 1.4) represents the number of hits found by the given number of search engines. For example, the by 1 (71) slice represents the number of unique hits found by one (and only one) search engine. Even with three Inktomi-based databases (iWon, MSN Search, and HotBot), there was not identical overlap between the three. However, the Inktomi databases are relatively similar.

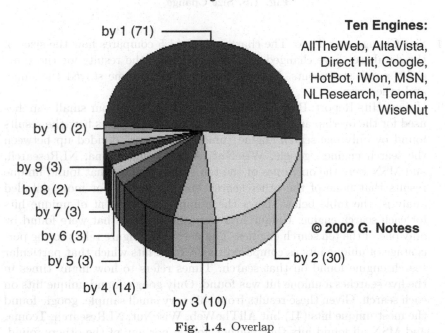

Overlap of 4 Searches
141 Pages on Mar. 6, 2002

by 1 (71)

Ten Engines:
AllTheWeb, AltaVista,
Direct Hit, Google,
HotBot, iWon, MSN,
NLResearch, Teoma,
WiseNut

by 10 (2)

by 9 (3)

by 8 (2)

by 7 (3)

by 6 (3)

© 2002 G. Notess

by 5 (3)

by 2 (30)

by 4 (14)

by 3 (10)

Fig. 1.4. Overlap

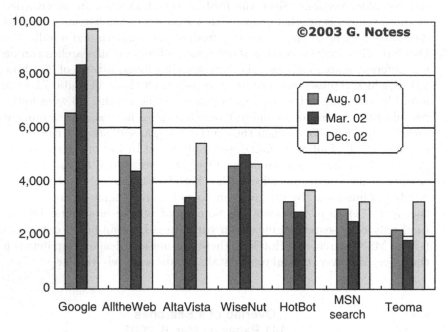

Fig. 1.5. Size Change

4. Size Change over Time: The chart in figure 1.5 compares how the sizes of the databases have changed over time, showing the results for the same eight search statements. Note how some shrank, some stayed the same, and other grew.

5. Unique Hits Report: Based on an analysis of the same four small searches used for the overlap analysis, the chart in figure 1.6 shows how the results found by only one search engine (unique hits) were divided up between the search engines. google, WiseNut, AllTheWeb, Teoma, NLResearch, and MSN were the only ones of the ten search engines that found unique results that none of the other search engines found. For more detailed analysis, the table below shows the number and percent of unique hits for each search engine. Unique hits were those URLs that were found by only one of the ten search engines. The percentage figure refers to the percentage of unique hits as compared to the total hits which that particular search engine found on that search. Times refers to how many times in the five searches a unique hit was found. Only google found unique hits on each search. Given these results from this very small sample, google found the most unique hits (41) but AllTheWeb, WiseNut, NLResearch, Teoma, and MSN all found hits that neither google nor any of the others found.

Distribution of 71 Unique Hits
by Search Engine

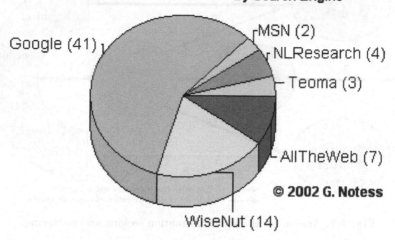

Fig. 1.6. Unique Hits

This example is from a total of 334 hits from four searches found by ten search engines. Of the 334 total hits, only 141 represented non-duplicative Web pages. And of those, 71 were unique hits in that they were only found by one of the ten search engines.

1.4 Users

As well expounded in *Johnson* [61], it is an almost universal finding in studies investigating human information behavior that people choose other people as their preferred source of information. Studies of academic researchers in both the sciences and the humanities have revealed the importance of consulting with colleagues at different stages of their research (*Ellis*, [33]). Professionals, such as engineers, nurses, physicians and dentists rely on co-workers and knowledgeable colleagues in their search for work-related information (Leckie et al. [74]). The explanation for the use of people as information sources has often been that they are 'typically easier and more readily accessible than the most authoritative printed sources' (*Case* [17], p. 142). The use of people is a least effort option in the search for information and, therefore, may not be the best sources available. The theory of social capital, however, suggests that the use of people as information sources is not necessarily an easy option, but may also require a considerable effort. The findings do suggest, however, that the choice of a person as a source of information is complex and needs to be investigated at a more finely-grained level to determine what factors affect who is chosen and under what circumstances (see chapter 6). The

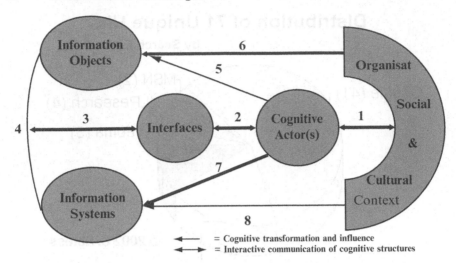

Fig. 1.7. Analytical Model of information seeking and retrieving

theory of social capital provides a useful framework within which to examine information behavior and to understand how social structure affects choice. Figure 1.7 presents a general analytical model of information seeking and retrieval proposed in *Ingwersen & Jarvelin* (forthcoming, [59]). It shows various players and their interaction in context in the field of information seeking and retrieval. We may observe cognitive actors such as information seekers in the middle surrounded by several kinds of contexts. These are formed by the actors' social, organizational and cultural affiliations, information objects, information systems and the interfaces for using them. The context for any node in the diagram consists of all the other nodes. For example, algorithmic and interactive information retrieval processes do not stand alone, but are special cases of information seeking behavior. Algorithmic information retrieval, that is, the interaction between information objects and algorithms, arrow (4), has no real meaning without human interaction with information retrieval systems, arrows (2-3). Interactive information retrieval itself functions as a special case of information seeking. Today, all information seeking becomes increasingly technology-driven because progressively more and more informal communication channels, such as mail, become formalized in systems because of the digitalization of communication. Information seeking behavior means the acquisition of information from knowledge sources. For instance, one may ask a colleague, through (in)formal channels such as social interaction in context (arrow 1), or through an information system (arrows 2-4). Secondly, actors operate in a dual context: that of information systems and information spaces surrounding them, and the socio-cultural-organizational context to the right. Over time, the latter context influences and, to a large degree, creates the information object space on the one hand (arrow 6) and the information technology infrastructure (arrow 8) on the other. In different roles, the actors themselves are not just information seekers but also authors of information objects (arrow 5) and creators of systems (arrow 7). Often, from the actor's point-of-view a proper system for analyzing information access may be a system of activities: work task, leisure interests and the social organization of these activities. We may take a look at augmenting task performance. After all, augmenting task performance is an important goal in working life and also in leisure activities. In the former it may be related to effectiveness and efficiency, and in the latter, if not efficiency, then at least quality.

Engelbart [34] suggested a conceptual framework for augmenting human intellect. Figures 1.8 shows a new way of looking at it and present an exemplary means-end hierarchy focusing on information seeking.

At the top of figure8 we have an actor's work task in context. In order to augment the actor's performance, one may improve her tools, her knowledge,

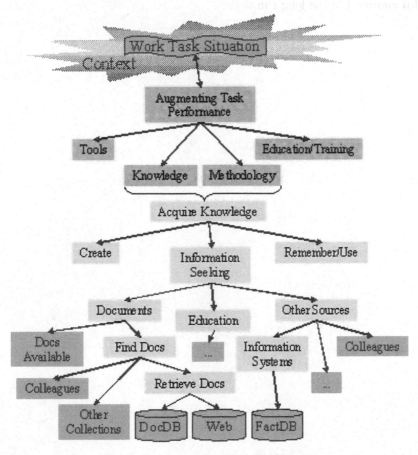

Fig. 1.8. Augmenting Task Performance

her methods and/or her training in the application of the former. Improving one's knowledge and methods means acquiring new knowledge, which can be done by creating it, remembering it or through information seeking. Information seeking, again, may reach towards education, documents or other sources. If one chooses documents, these may be documents immediately to hand, documents from colleagues or nearby collections, or documents retrieved from various databases through searching. If one chooses other sources than documents, these may be factual databases, colleagues, etc. In this figure, information seeking and its manifestations may seem somewhat remote from the work task - with document retrieval even more remote and behind many decisions. Nevertheless, our view is that information seeking and retrieval belongs to a context in real life, such as the work task context. The distance does not make information seeking and retrieval independent of work tasks or other activities; it needs to contribute to them. This sets many requirements on augmentation through information seeking and retrieval if it is going to be useful and used in the long run.

2

Web Graphs

2.1 Introduction

A Web page is dynamic and bi-directional. It points to other Web links as
well as other Web links point to it. A Web page gets updated and so new
Web links are discovered as new links are added. It is a living community of
links. The contextual amount of variability between two Web pages could be
very high. Not only they deal with different subjects, but they could be in
different languages, have nothing in common, no standards are imposed on
the content, and reflect subjective ideas than commonly established scientific
explorations. Human judgment on Web content is more subjective, and noisier
than in a citation. It is hard to keep a control on the quality of Web pages
that are created. Web pages serve many purposes not just to link 2 or more
Web pages. Hubs and authorities (*Kleinberg* [64]) have been developed to
better assess the influence that a Web page has on other Web pages that
link to it. Web pages also reflect the "signatures of intelligence" in our era
and contain rich information on our collective society as a whole. Citation
analysis is a very important field for information scientists for mining citations
in journals. The comparison above sheds some light on the parallel between
a Web page and a citation in a journal ([64]). This just to emphasize the
interest that Web mining experts ought to have to analyze Web links even
though it is a task in its infancy, and even though that the principles that
govern Web pages and Web links are different from those of scientific citations.
Web pages are not scrutinized like scientific journals are. New links in a Web
page can have an impact on the whole search done and the always changing
size of the Web can reflect patterns of indexing, crawling for search engine
designers that feed on these Web links (*Meghabghab* [90]). More importantly,
users need to be aware not only of the set of results returned, but also of
the set of results not returned or the percentage of "rejected Web pages"
for each query. To fully comprehend and assess the discovery of hubs and
authorities and the influence that hyperlinks between different Web pages
have over other Web pages, Graph Theory can be used to better understand

G. Meghabghab and A. Kandel: *Search Engines, Link Analysis, and User's Web Behavior*,
Studies in Computational Intelligence (SCI) **99**, 23–45 (2008)
www.springerlink.com

the measures developed in Kleinberg ([64]). Nodes represent static html pages and hyperlinks represent directed edges. But never an attempt have been made to study Web graphs in the (Out-degree, in-degree) coordinate space, neither has citation analysis on the Web has been applied in this new coordinate space and has revealed the complexity of citation analysis in a variety of Web graphs. Graph Theory can be used also to discover new patterns that appear in a citation graph. The same idea can be used to measure the distance between 2 Web pages. Measuring the topological structure richness of a collection of Web pages is an important aspect of Web pages never explored before and never applied before on hubs and authorities. The next section is a reminder on concepts borrowed from Graph Theory to help analyze hyperlinks, and the richness of the WWW as an information network of ideas and decisions.

2.2 Basic Graph Theory Applied to Web Pages

A graph is a directed link. A link on a Web page connects one document to another. A link is not only of a navigational nature but also can represent an endorsement to the target page. When we consider more than just one link, we could explore characteristics of the Web space. Spatial relations between Web pages can help understand the topology of a Web page and in general of the Web space. In the space of all Web pages W, let $A \in W$ to mean a page A belongs to the space of all Web pages. The Web page A represents a graph. In that graph, if there is a link to another Web page B, we can say that: A is related to B by the link. In symbolic terms we can write $(A, B) \in R$, where R is the relation "point to". We can add the following observations on the relation R :

1. If every Web page is related to itself, we say that R is reflexive.
2. For all Web pages X and Y that belong to W, if $(X, Y) \in R \to (Y, X) \in R$. Web pages X and Y in that case represent mutual endorsement. Then the relation is then said to be symmetric.
3. For all Web pages X and Y that belong to W, if $(X, Y) \in R \to (Y, X)$ $\notin R$. Web pages X and Y are linked in a unidirectional way. Then the relation is then said to be anti-symmetric.
4. For all Web pages that belong to W, when a Web page X cites another Web page Y and that last page cites another Web page Z, we can say that R is transitive:

$$(X, Y) \in R \text{ and } (Y, Z) \in R \to (X, Z) \in R$$

5. When a Web page X cites another Web page Y and Y does not cite X, X endorses Y and Y does not endorse X, we can say that R is not symmetric:

$$(X, Y) \in R \text{ but } (Y, X) \notin R.$$

6. When 2 Web pages X and Y point to a distinct 3rd Web page Z, then we could say that the 2 Web pages are related through a very special relationship similar to a filtering relationship or bibliographic coupling (*Kessler* [63]). This kind of relationship does not have a name in the algebraic properties of R.

$$(X, Z) \in R \text{ and } (Y, Z) \in R$$

7. Conversely when one Web page X points to 2 distinct Web pages Y and Z, then we say that X co-cites Y and Z. Co-citation is a term borrowed from the field of biblio-metric studies (*Small* [122]). Co-citation has been used as a measure of similarity between Web pages by *Larson* [71] and *Pitkow & Pirolli* [113]. *Small & Griffith* [124] used breadth-first search to compute the connected components of the uni-directed graphs in which 2 nodes are joined by an edge if and only if they have a positive co-citation value. This kind of relationship does not have a name in the algebraic properties of R.

$$(X, Y) \in R \text{ and } (X, Z) \in R.$$

These 7 properties are the simplest common patterns that can be perceived on the Web. These 7 properties can blend together to form more complex patterns that are indicative of emerging links or communities on the Web. These complex patterns can model properties of Web pages that can be qualified as "authoritative Web pages". An authoritative Web page is a Web page that is pointed at by many other Web pages. Other emerging complex patterns can model Web pages that can be qualified as survey Web pages or "hub Web pages" because they cite "authoritative Web pages".

2.2.1 Adjacency Matrix Representation of a Web Graph

To further apply all these properties, consider a directed graph G that represents hyperlinks between Web pages. Consider also its adjacency matrix A. An entry a_{pq} is defined by the following:

$$a_{pq} = 1 \quad \text{if there is a link between Web pages p and q.}$$
$$= 0 \quad \text{Otherwise}$$

Here some of the properties that could be discovered from an adjacency matrix perspective:

1. A graph is said to be reflexive if every node in a graph is connected back to itself, i.e., $a_{pp} = 1$. The situation will happen if a page points back to itself.
2. A graph is said to be symmetric if for all edges p and q in X: iff $a_{pq} = 1$ then $a_{qp} = 1$. We say in this case that there is mutual endorsement.

3. A graph is said to be not symmetric if there exists 2 edges p and q in G such that iff $a_{pq} = 1$ then $a_{qp} = 0$. We say in this case that there is endorsement in one direction.

4. A graph is said to be transitive if for all edges p,q, and r: Iff $a_{pq} = 1$ and $a_{qr} = 1$ then $a_{pr} = 1$ We say in this case that all links p endorse links r even though not directly.

5. A graph is said to be anti-symmetric iff for all edges p and q: Iff $a_{pq} = 1$ then $a_{qp} = 0$

6. If 2 different Web pages p and q point to another Web page r then we say that there is social filtering. This means that these Web pages are related through a meaningful relationship. $a_{pr} = 1$ and $a_{qr} = 1$

7. If a single page p points to 2 different Web pages q and r then we say that there is co-citation. $a_{pq} = 1$ and $a_{pr} = 1$

Now consider 2 linear transformations defined on unit vectors a and h as follows:

$$a = A^T h \tag{2.1}$$

and:

$$h = Ah \tag{2.2}$$

thus:

$$a = AA^T a \tag{2.3}$$

and:

$$h = A^T Ah \tag{2.4}$$

By examining closely the entries of these product matrices AA^T and $A^T A$, we find that 2 matrices are symmetric. with the following properties observed:

1. An entry (p, p) in AA^T means the number of Web pages that come out of p. We call that number the out-degree or o_d.

2. An entry (p, p) in $A^T A$ means the number of Web pages that point towards p. We call that number in-degree or i_d.

3. An entry (p, q) in $A^T A$ represents the number of Web pages that are in common between p and q that point towards p and q.

4. An entry (p, q) in AA^T represents the number of Web pages that came out of p and q that are in common.

To illustrate the above points let us look at the following graph G. Here are the algebraic properties of R in G:

1. R is not reflexive.
2. R is not symmetric.
3. R is not transitive.

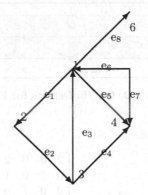

Fig. 2.1. A General Graph

4. R is anti-symmetric.
5. (1,4) ∈ R, (3,4) ∈ R, and (5,4) ∈ R: we say that the vertex with the highest number of Web pages pointing to it.
6. (5,1) ∈ R and (5,6) ∈ R: 5 co-cites 1 and 6.

Here is the corresponding adjacency matrix for figure 2.1.

$$O_d$$

$$
\mathbf{A} =
\begin{matrix} 3 \\ 1 \\ 2 \\ 0 \\ 2 \\ 0 \end{matrix}
\begin{bmatrix}
0 & 1 & 0 & 1 & 0 & 1 \\
0 & 0 & 1 & 0 & 0 & 0 \\
1 & 0 & 0 & 1 & 0 & 0 \\
0 & 0 & 0 & 0 & 0 & 0 \\
1 & 0 & 0 & 1 & 0 & 0 \\
0 & 0 & 0 & 0 & 0 & 0
\end{bmatrix}
$$

Building A^T *yields:*

$$
\mathbf{A^T} =
\begin{bmatrix}
0 & 0 & 1 & 0 & 1 & 0 \\
1 & 0 & 0 & 0 & 0 & 0 \\
0 & 1 & 0 & 0 & 0 & 0 \\
1 & 0 & 1 & 0 & 1 & 0 \\
0 & 0 & 0 & 0 & 0 & 0 \\
1 & 0 & 0 & 0 & 0 & 0
\end{bmatrix}
$$

Building $C = AA^T$ yields:

$$
\mathbf{C} = \mathbf{AA^T} =
\begin{bmatrix}
3 & 0 & 1 & 0 & 0 & 0 \\
0 & 1 & 0 & 0 & 0 & 0 \\
1 & 0 & 2 & 0 & 2 & 0 \\
0 & 0 & 0 & 0 & 0 & 0 \\
0 & 0 & 2 & 0 & 2 & 0 \\
1 & 0 & 0 & 0 & 0 & 0
\end{bmatrix}
$$

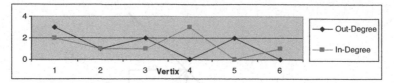

Fig. 2.2. Out/In Degrees for Figure 2.1

Building $D = A^T A$ yields:

$$\mathbf{D = A^T A} = \begin{bmatrix} 2 & 0 & 0 & 2 & 0 & 0 \\ 0 & 1 & 0 & 1 & 0 & 1 \\ 0 & 0 & 1 & 0 & 0 & 0 \\ 2 & 1 & 0 & 3 & 0 & 1 \\ 0 & 0 & 0 & 0 & 0 & 0 \\ 0 & 1 & 0 & 1 & 0 & 1 \end{bmatrix}$$

The next figure illustrates the in-degrees and out degrees for the graph G (see figure 2.2):

How far away are 2 Web pages in a Web graph? The adjacency matrix A can be used to calculate the length of the path than can separate 2 distinct Web pages. To further explore such an idea, consider the power matrices of A, i.e., A^2, A^3, A^4, A^n for a graph of n vertices. If we calculate the value of A^2 for $Graph_1$ we have:

$$\mathbf{A^2} = \begin{bmatrix} 0 & 0 & 1 & 0 & 0 & 0 \\ 1 & 0 & 0 & 1 & 0 & 1 \\ 0 & 1 & 1 & 1 & 0 & 1 \\ 0 & 0 & 0 & 0 & 0 & 1 \\ 0 & 1 & 0 & 1 & 0 & 1 \\ 0 & 0 & 0 & 0 & 0 & 0 \end{bmatrix}$$

Every non-zero element in A^2 means that to travel from vertex i to vertex j we will need 2 Web links to get there. Thus considering that $A^2(2,4) = 1$ means that the distance from vertex 2 to vertex 4 is 2. This can be verified on Figure 2.1 where (2,3) $\in$ R and (3,4) $\in$ R If we calculate the rest of powers of A, i.e., $A^3, A^4, ...A^n$, and we reach a value $m < n$ such that $A^m = A$ then we say that any 2 Web pages in that graph are m pages or clicks away. Applying this to $Graph_1$, one can see that $A^4 = A$. This means that for any 2 Web pages, the furthest away they are, is 4 Web pages. An example in $Graph_1$ is Web page 1 and 6, where to reach 6 from 1 we can travel directly with a path of length 1 or through vertices 2, 3, 1, 6. Expanding the idea of the distances of Web pages over the WWW, Albert et al. [3] were able to show that the distribution of Web pages over the Web constitutes a power law and that the distance between far away connected Web pages is 19. In other words, to move along the whole WWW, it will take approximately 19 Web pages or clicks at most. Thus the diameter of the WWW is 19 clicks.

2.2.2 Incidence Matrix Representation of a Web Graph

Another interesting representation of a graph is an incidence matrix. Let us consider the same graph G in Figure 2.1 and its corresponding incidence matrix I. To build the incidence matrix of a graph we label the rows with the vertices and the columns with the edges (in any order). The entry i_{ve} for row r (vertex v) and column c (edge e) is such:

$$
\begin{aligned}
i_{ve} &= 1 && \text{if e is incident to v} \\
&= 0 && \text{Otherwise}
\end{aligned}
\tag{2.5}
$$

Notice that i_d, which is the number of edges incident on a vertex v, can be deduced from the incidence matrix. We also added to I the row s which the sum of all the values in a given column.

$$
I = \begin{array}{c} \\ 1 \\ 2 \\ 3 \\ 4 \\ 5 \\ 6 \\ s \end{array}
\begin{array}{c}
\begin{array}{cccccccc} e_1 & e_2 & e_3 & e_4 & e_5 & e_6 & e_7 & e_8 \end{array} \\
\left[\begin{array}{cccccccc}
0 & 0 & 1 & 0 & 0 & 0 & 1 & 0 \\
1 & 0 & 0 & 0 & 0 & 0 & 0 & 0 \\
0 & 1 & 1 & 0 & 0 & 0 & 0 & 0 \\
0 & 0 & 0 & 0 & 1 & 1 & 0 & 0 \\
0 & 0 & 0 & 0 & 0 & 0 & 0 & 1 \\
0 & 0 & 0 & 0 & 0 & 0 & 0 & 1 \\
1 & 1 & 1 & 1 & 1 & 1 & 1 & 1
\end{array}\right]
\end{array}
\begin{array}{c} i_d \\ 2 \\ 1 \\ 1 \\ 3 \\ 0 \\ 1 \\ 7 \end{array}
$$

We could deduce from I that Web page 4 or vertex 4 is the one with the highest incidence of links to it. Web page 4 is an authoritative Web page since it is the Web page with most links pointing to it. We can deduce through the value of s that there are no bi-directional links on this graph. That is why this graph is anti-symmetric. One way to look at how matrix A and vector i_d relate is by considering the following matrix-vector multiplication $A^T U$ where A^T is the transpose of A already computed in 2.2.1 and U is the Unit vector 1. Applying $A^T U$ to $Graph_1$ results in the following:

$$
A^T \times U = i_d
\tag{2.6}
$$

The matrix I has been ignored in the literature on Web Graph Theory ([64]). Looking closely at II^T yields the following matrix:

$$
II^T = \begin{bmatrix}
2 & 0 & 0 & 0 & 0 & 0 \\
0 & 1 & 0 & 0 & 0 & 0 \\
0 & 0 & 1 & 0 & 0 & 0 \\
0 & 0 & 0 & 3 & 0 & 0 \\
0 & 0 & 0 & 0 & 0 & 0 \\
0 & 0 & 0 & 0 & 0 & 1
\end{bmatrix}
$$

II^T is a diagonal matrix. Its eigenvalues vector is equal to the in-degree vector i_d. Its eigenvectors constitute the columns of the unit matrix of size 6×6:

$$Eigenvalue(II^T) = i_d \qquad (2.7)$$

By looking at equations 2.6 and 2.7 we conclude that:

$$Eigenvalue(II^T) = i_d = A^T U \qquad (2.8)$$

2.2.3 Bipartite Graphs

A Bipartite graph G is a graph where the set of vertices can be divided into sets V_1 and V_2 such that each edge is incident on one vertex in V_1 and one vertex in V_2. Graph G in Figure 2.1 is not an actual Bipartite graph. To make G an actual bipartite graph, a possible bipartite graph G_1 can be designed. If we let $V_1 = \{1, 3, 5\}$ and $V_2 = \{2, 4, 6\}$, then we can take out the 2 edges e_3 and e_7 that were in and then the new graph G_1 will become a bipartite graph as in Figure 2.3. In other related works, tree structures have been used to design better hyperlink structures (*Botafogo et al.* [13]). The reverse process of discovering tree structures from hyperlink Web pages and discover hierarchical structures has also been studied (*Mukherjea et al.* [103]; *Pirolli et al.* [112]). In case of topic search, we do not need to extract a Web structure from the Web. Often the user is interested in finding a small number of authoritative pages on the search topic. These pages will play an important role in a tree had we extracted the tree structure itself. An alternative to extracting trees from a Web graph is to use a ranking method to the nodes of the Web graph. In this section we review such methods proposed in the literature. Some basic concepts have to be laid down before doing that. We conclude that the topology of the links in Web pages affect search performance and strategies of the WWW.

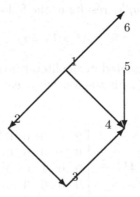

Fig. 2.3. A Bipartite Graph G_1

2.2.4 Web Topology

In this section, we will explore how different can Web graphs be? Can we classify these different Web pages? How complex these Web pages can appear to be? Different categories of Web graphs can emerge according to their o_d value and i_d value or what we defined in section 2.1 as Out-degree and in-degree correspondingly. Even though we do not pretend that this classification is exhaustive, but different kinds of graphs were gathered to represent the possible different variety of Web pages. Emerging Web graphs can be complex and rich in structure and links more than Web page designers do realize.

In-degree Web Graphs

Complex pages can emerge with large in-degree that looks like Figure 2.4: Figure 2.5 illustrates such in-degree Web pages by looking at their In/Out Degree chart:

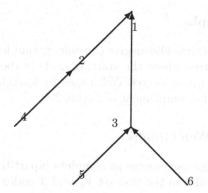

Fig. 2.4. An In-Degree Graph

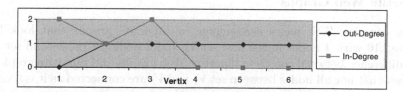

Fig. 2.5. Out/In Degrees for Figure 2.4

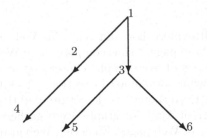

Fig. 2.6. An Out-Degree Graph

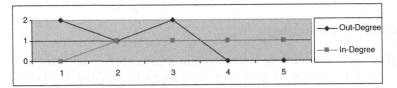

Fig. 2.7. Out/In Degrees for Figure 2.6

Out-degree Web Graphs

Complex Web pages with large Out-degree can emerge that look like Figure 2.6. Such graph becomes a tree where the starting node is the root of the tree: Figure 2.7 illustrates such Out-degree Web pages by looking at their In/Out Degree chart, which is the complement of Figure 2.5:

Complete Bipartite Web Graphs

Other complex Web pages can emerge as complete bipartite graphs that look like Figure 2.8 with 3 nodes in the first set V_1 and 3 nodes in the second set V_2: Remember that the topology of complete bipartite graphs like the one in Figure 2.8 is unique. Figure 2.9 illustrates such Complete Bipartite Web pages by looking at their In/Out Degree chart.

Bipartite Web Graphs

Other complex Web pages can emerge as bipartite graphs that look like Figure 2.10 with 4 nodes in the first set V_1 and 2 nodes in the second set V_2: The difference between complete bipartite Web graphs and bipartite graphs is the fact that not all nodes between set V_1 and V_2 are connected as it is seen in Figure 2.10: pages with large in-degree or Out-degree play an important role in Web algorithms in general. Section 3.4 will apply Kleinberg's algorithm [64] on these different graphs. Figure 2.11 illustrates such Bipartite Web pages by looking at their IN/Out Degree chart.

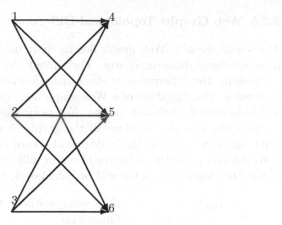

Fig. 2.8. A Complete Bipartite Graph

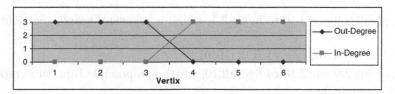

Fig. 2.9. Out/In Degrees for Figure 2.8

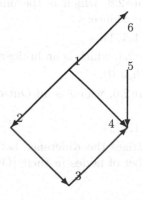

Fig. 2.10. A Bipartite Graph G_1

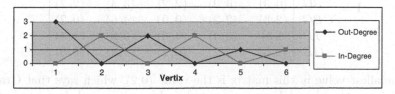

Fig. 2.11. Out/In Degrees for Figure 2.10

2.2.5 Web Graphs Topological Difference

The signature of a Web graph lies in their in/out degree as it can be seen from all these different charts. The in/out degree of a graph can be used to measure the differences or similarities between different topological Web structures. The signature of a Web graph lies in that. The Euclidean distance will help classify different graphs. Not only will be interested in the size of a graph but also the structure of the graph. A graph will be assumed to be symbolized by 2 vectors the in-degree id vector and the Out-degree od vector. Next the average of the in-degree id vector will be calculated, the average value of the Out-degree od vector will be calculated, and say that for a vertex v:

$$
\begin{aligned}
o_d(v) &= 1 &&\text{if its value is above the average value} \\
&= 0 &&\text{Otherwise}
\end{aligned} \tag{2.9}
$$

$$
\begin{aligned}
i_d(v) &= 1 &&\text{if its value is above the average value} \\
&= 0 &&\text{Otherwise}
\end{aligned} \tag{2.10}
$$

By applying 2.9 and 2.10 to figure 2.1, which is a general graph, we could deduce:
$o_d = (1, 0, 1, 0, 1, 0)$, $i_d = (1, 0, 0, 1, 0, 0)$.

By applying 2.9 and 2.10 to figure 2.10, which is a possible Bipartite graph of figure 2.1, we deduce:
$o_d = (1, 0, 1, 0, 1, 0)$, $i_d = (0, 1, 0, 1, 0, 1)$.

By applying 2.9 and 2.10 to figure 2.8, which is the only complete possible Bipartite graph on figure 2.10, we deduce:
$o_d = (1, 1, 1, 0, 0, 0)$, $i_d = (0, 0, 0, 1, 1, 1)$.

By applying 2.9 and 2.10 to figure 2.4, which is an in-degree graph, we deduce:
$o_d = (0, 1, 1, 1, 1, 1)$, $i_d = (1, 1, 1, 0, 0, 0)$.

By applying 2.9 and 2.10 to figure 2.6, which is an Out-degree graph, we deduce:
$o_d = (1, 1, 1, 0, 0, 0)$, $i_d = (0, 1, 1, 1, 1, 1)$.

The following matrix M summarizes the difference between these different Web graphs with the same number of nodes in their (Out-degree, in-degree) coordinate space:

$$
\mathbf{I} =
\begin{array}{c}
\\ GG \\ BG \\ CBG \\ ID \\ OD
\end{array}
\begin{array}{c}
GG \\
\left[\begin{array}{c} (0,0) \\ (0,3) \\ (2,3) \\ 3,3) \\ (2,5) \end{array}\right.
\end{array}
\begin{array}{c}
BG \\
(0,2) \\ (0,0) \\ (2,2) \\ (3,4) \\ (2,2)
\end{array}
\begin{array}{c}
CBG \\
(2,3) \\ (2,2) \\ (0,0) \\ (4,6) \\ (0,2)
\end{array}
\begin{array}{c}
ID \\
(3,3) \\ (3,4) \\ (4,6) \\ (0,0) \\ (4,4)
\end{array}
\begin{array}{c}
OD \\
\left.\begin{array}{c} (2,5) \\ (2,2) \\ (0,2) \\ (4,4) \\ (0,0) \end{array}\right]
\end{array}
$$

The smallest value in this matrix is the value (0,2), which says that Graphs 8 and 6 are the closest because a complete bipartite graph is a form of an

Out-degree graph with many roots. The next best smallest value in the matrix M is (0,3) which says that general graphs and bipartite graphs are the closest among all other graphs. The largest value in M is (4,6) which says that complete bipartite graphs and in-degree are the farthest apart. The next biggest difference is (4,4) which says that the next largest difference is between in-degrees trees and Out-degree trees, which is evident form the structure of the trees. It also shows that bipartite graphs are as close to Out-degree trees and complete bipartite graphs than in-degree trees which can be concluded from the statement before. We conclude that in the coordinate space of (Out-degree, In-Degree) the following metric of graphs topology stands:

$$|(CBG)| < |(OD)| < |(BG)| < |(GG)| < |(ID)| \qquad (2.11)$$

Where   CBG = complete   bipartite   graph,   OD = Out-degree   Trees,
BG = bipartite graph, GG = General Graphs, and ID = In-Degree Trees.

2.3 Bow Tie Graphs

2.3.1 Introduction

Broder et al.'s [15] study analyzed the connectivity of more than 1.5 billion links on over 200 million Web pages. By creating a map showing all links pointing to and from this large sample of pages, the researchers found a comprehensive topography of the Web.

We now give a more detailed description of the structure in Figure 2.12 which represents connectivity of the Web. One can pass from any node of IN through SCC to any node of OUT. Hanging off IN and OUT are TENDRILS containing nodes that are reachable from portions of IN, or that can reach portions of OUT, without passage through SCC. It is possible for a TENDRIL hanging off from IN to be hooked into a TENDRIL leading into OUT, forming a TUBE – a passage from a portion of IN to a portion of OUT without touching SCC. The sizes of the various components are as seen in table 2.1

Unlike previous models portraying the Web as clusters of sites forming a well-connected sphere, the results showed that the Web's structure more closely resembles a bow tie consisting of three major regions (a knot and two bows), and a fourth, smaller region of pages that are disconnected from the basic bow-tie structure.

1. At the center of the bow tie is the knot, which the study calls the strongly connected core. This core is the heart of the Web. pages within the core are strongly connected by extensive cross-linking among and between themselves. Links on core pages enable Web surfers to travel relatively easily from page to page within the core.
2. left bow consists of origination pages that eventually allow users to reach the core, but that cannot themselves be reached from the core. Origination

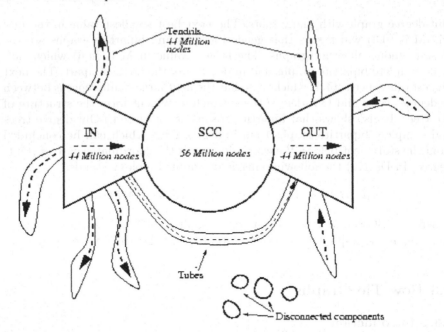

Fig. 2.12. Bow Tie Graphs

Table 2.1. Size vs. Region in a Bow Tie graph

Region	SCC	IN	OUT	TENDRILS	DISC.	Total
Size	$56 * 10^6$	$43 * 10^6$	$43 * 10^6$	$43 * 10^6$	$16 * 10^6$	$203 * 10^6$

pages are typically new or obscure Web pages that haven't yet attracted interest from the Web community (they have no links pointing to them from pages within the core), or that are only linked to from other origination pages. Relatively closed community sites such as Geocities and Tripod are rife with origination pages that often link to one another but are seldom linked to from pages within the core.

3. The right bow consists of "termination" pages that can be accessed via links from the core but that do not link back into the core. Many commercial sites don't point anywhere except to themselves. Instead, corporate sites exist to provide information, sell goods or services, and otherwise serve as destinations in themselves, and there is little motivation to have them link back to the core of the Web.

2.3.2 Parametric Characteristics of Bow Tie Graphs

Parametric equations of the form $x = \cos(at)$ and $y = \sin(bt)$, where a and b are constants, occur in electrical theory. The variables x and y could represent voltages or currents at time t. The resulting curve is often difficult to sketch,

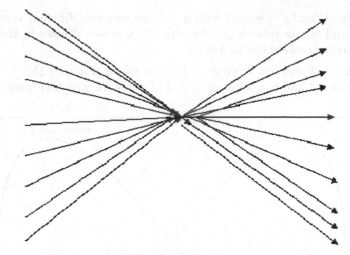

Fig. 2.13. A 10*1*10 Bow Tie Graph

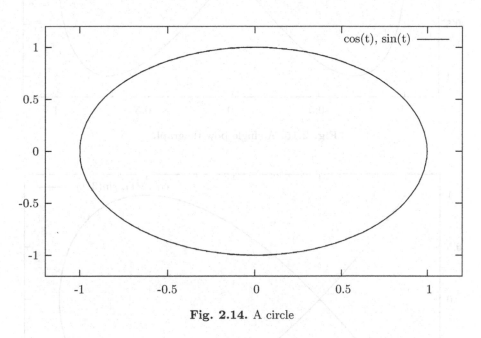

Fig. 2.14. A circle

but the graph can be represented on the screen of an oscilloscope when voltages or currents are imposed on the input terminals. These figures are called Lissajous curves. The times will be restricted to t = 0 to t = 2*π. Here are different cases:

1. For $x = \cos(at)$ and $y = \sin(bt)$ with a = b, $x = a\cos(t)$ and $y = a\sin(t)$, the graph will be always a circle. (Figure 2.14).

2. $x = \cos(at)$ and $y = \sin(at)$ with $a < b$, we consider different values of a and b and obtain bow tie graphs with the number of bows in the graph equal to the ratio of b/a as follows:

(a) For a = 1 and b = 2, we get a bow tie graph (Figure 2.15).
(b) For a = 2 and b = 3, we get a 2/3 of a bow tie graph (Figure 2.16).

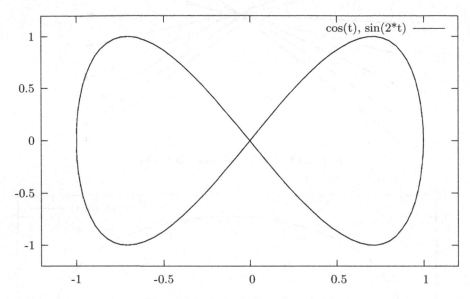

Fig. 2.15. A single bow tie graph

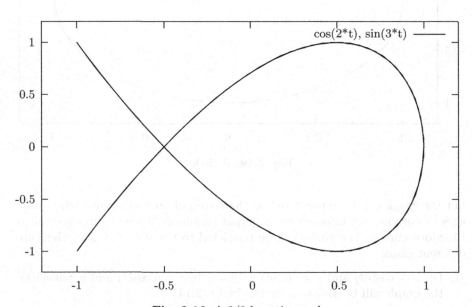

Fig. 2.16. A 2/3 bow tie graph

(c) For a = 1 when b = 4, we get a double bow tie or 4 bows (Figure 2.17).
(d) For a = 1 when b = 5 we get 5 bows (Figure 2.18).
(e) For a = 1 and b = 8, we get 8 bows (Figure 2.19).

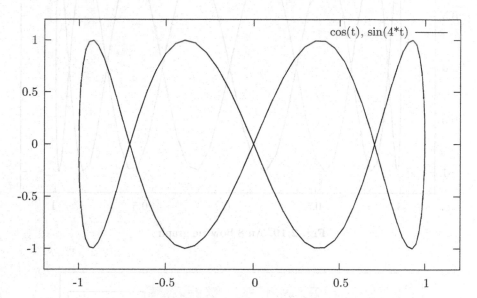

Fig. 2.17. A double bow tie graph

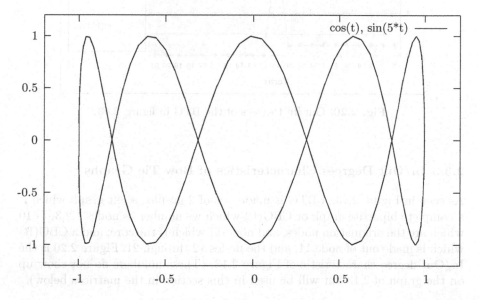

Fig. 2.18. A 5 bow tie graph

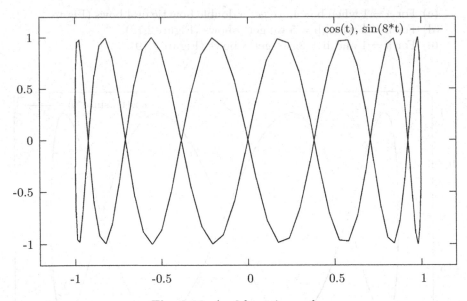

Fig. 2.19. An 8 bow tie graph

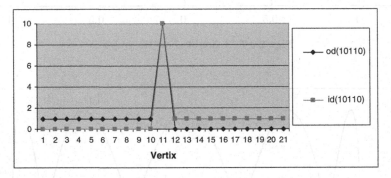

Fig. 2.20. Out/In Degrees of the BTG in figure 2.13

2.3.3 In/Out Degrees Characteristics of Bow Tie Graphs

As seen in Figure 2.13, a BTG is made out of 2 graphs, a left graph which is a complete bipartite graph or CBG(L) which we number as nodes 1,2,3,···,10 which are the origination nodes, and node 11 which is the core; and a CBG(R) which is made out of node 11, and the nodes 12 through 21. Figure 2.20 is the In/Out degree characteristic of Figure 2.13. (These numbers do not show up on the graph of 2.13 but will be used in this section in the matrices below).

$$0_d$$

$$
\begin{bmatrix}
0 & 0 & 0 & 0 & 0 & 0 & 0 & 0 & 0 & 0 & 0 & 1 & 0 & 0 & 0 & 0 & 0 & 0 & 0 & 0 & 0 & 0 \\
0 & 0 & 0 & 0 & 0 & 0 & 0 & 0 & 0 & 0 & 0 & 1 & 0 & 0 & 0 & 0 & 0 & 0 & 0 & 0 & 0 & 0 \\
0 & 0 & 0 & 0 & 0 & 0 & 0 & 0 & 0 & 0 & 0 & 1 & 0 & 0 & 0 & 0 & 0 & 0 & 0 & 0 & 0 & 0 \\
0 & 0 & 0 & 0 & 0 & 0 & 0 & 0 & 0 & 0 & 0 & 1 & 0 & 0 & 0 & 0 & 0 & 0 & 0 & 0 & 0 & 0 \\
0 & 0 & 0 & 0 & 0 & 0 & 0 & 0 & 0 & 0 & 0 & 1 & 0 & 0 & 0 & 0 & 0 & 0 & 0 & 0 & 0 & 0 \\
0 & 0 & 0 & 0 & 0 & 0 & 0 & 0 & 0 & 0 & 0 & 1 & 0 & 0 & 0 & 0 & 0 & 0 & 0 & 0 & 0 & 0 \\
0 & 0 & 0 & 0 & 0 & 0 & 0 & 0 & 0 & 0 & 0 & 1 & 0 & 0 & 0 & 0 & 0 & 0 & 0 & 0 & 0 & 0 \\
0 & 0 & 0 & 0 & 0 & 0 & 0 & 0 & 0 & 0 & 0 & 1 & 0 & 0 & 0 & 0 & 0 & 0 & 0 & 0 & 0 & 0 \\
0 & 0 & 0 & 0 & 0 & 0 & 0 & 0 & 0 & 0 & 0 & 1 & 0 & 0 & 0 & 0 & 0 & 0 & 0 & 0 & 0 & 0 \\
0 & 0 & 0 & 0 & 0 & 0 & 0 & 0 & 0 & 0 & 0 & 1 & 0 & 0 & 0 & 0 & 0 & 0 & 0 & 0 & 0 & 0 \\
0 & 0 & 0 & 0 & 0 & 0 & 0 & 0 & 0 & 0 & 0 & 0 & 1 & 1 & 1 & 1 & 1 & 1 & 1 & 1 & 1 & 1 \\
0 & 0 & 0 & 0 & 0 & 0 & 0 & 0 & 0 & 0 & 0 & 0 & 0 & 0 & 0 & 0 & 0 & 0 & 0 & 0 & 0 & 0 \\
0 & 0 & 0 & 0 & 0 & 0 & 0 & 0 & 0 & 0 & 0 & 0 & 0 & 0 & 0 & 0 & 0 & 0 & 0 & 0 & 0 & 0 \\
0 & 0 & 0 & 0 & 0 & 0 & 0 & 0 & 0 & 0 & 0 & 0 & 0 & 0 & 0 & 0 & 0 & 0 & 0 & 0 & 0 & 0 \\
0 & 0 & 0 & 0 & 0 & 0 & 0 & 0 & 0 & 0 & 0 & 0 & 0 & 0 & 0 & 0 & 0 & 0 & 0 & 0 & 0 & 0 \\
0 & 0 & 0 & 0 & 0 & 0 & 0 & 0 & 0 & 0 & 0 & 0 & 0 & 0 & 0 & 0 & 0 & 0 & 0 & 0 & 0 & 0 \\
0 & 0 & 0 & 0 & 0 & 0 & 0 & 0 & 0 & 0 & 0 & 0 & 0 & 0 & 0 & 0 & 0 & 0 & 0 & 0 & 0 & 0 \\
0 & 0 & 0 & 0 & 0 & 0 & 0 & 0 & 0 & 0 & 0 & 0 & 0 & 0 & 0 & 0 & 0 & 0 & 0 & 0 & 0 & 0 \\
0 & 0 & 0 & 0 & 0 & 0 & 0 & 0 & 0 & 0 & 0 & 0 & 0 & 0 & 0 & 0 & 0 & 0 & 0 & 0 & 0 & 0 \\
0 & 0 & 0 & 0 & 0 & 0 & 0 & 0 & 0 & 0 & 0 & 0 & 0 & 0 & 0 & 0 & 0 & 0 & 0 & 0 & 0 & 0 \\
0 & 0 & 0 & 0 & 0 & 0 & 0 & 0 & 0 & 0 & 0 & 0 & 0 & 0 & 0 & 0 & 0 & 0 & 0 & 0 & 0 & 0
\end{bmatrix}
\begin{matrix}
1 \\ 1 \\ 1 \\ 1 \\ 1 \\ 1 \\ 1 \\ 1 \\ 1 \\ 1 \\ 10 \\ 1 \\ 1 \\ 1 \\ 1 \\ 1 \\ 1 \\ 1 \\ 1 \\ 1 \\ 1
\end{matrix}
$$

The above matrix is the adjacency matrix $p_{10*1*10}$ of figure 2.13. p_{10x1} is the adjacency matrix of figure 2.21.

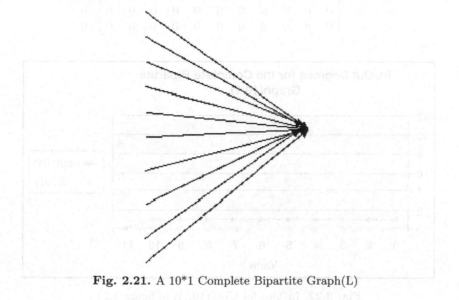

Fig. 2.21. A 10*1 Complete Bipartite Graph(L)

$$0_d$$

$$
p_{10x1} = \begin{bmatrix}
0 & 0 & 0 & 0 & 0 & 0 & 0 & 0 & 0 & 0 & 1 \\
0 & 0 & 0 & 0 & 0 & 0 & 0 & 0 & 0 & 0 & 1 \\
0 & 0 & 0 & 0 & 0 & 0 & 0 & 0 & 0 & 0 & 1 \\
0 & 0 & 0 & 0 & 0 & 0 & 0 & 0 & 0 & 0 & 1 \\
0 & 0 & 0 & 0 & 0 & 0 & 0 & 0 & 0 & 0 & 1 \\
0 & 0 & 0 & 0 & 0 & 0 & 0 & 0 & 0 & 0 & 1 \\
0 & 0 & 0 & 0 & 0 & 0 & 0 & 0 & 0 & 0 & 1 \\
0 & 0 & 0 & 0 & 0 & 0 & 0 & 0 & 0 & 0 & 1 \\
0 & 0 & 0 & 0 & 0 & 0 & 0 & 0 & 0 & 0 & 1 \\
0 & 0 & 0 & 0 & 0 & 0 & 0 & 0 & 0 & 0 & 1 \\
0 & 0 & 0 & 0 & 0 & 0 & 0 & 0 & 0 & 0 & 0 \\
\end{bmatrix}
\begin{matrix}
1 \\ 1 \\ 1 \\ 1 \\ 1 \\ 1 \\ 1 \\ 1 \\ 1 \\ 1 \\ 0
\end{matrix}
$$

The CBG(L) looks like (See Figure 2.21) and its In/Out degree characteristic in Figure 2.22. p_{1x10} is the adjacency matrix of figure 2.23.

$$0_d$$

$$
p_{1x10} = \begin{bmatrix}
0 & 1 & 1 & 1 & 1 & 1 & 1 & 1 & 1 & 1 & 1 \\
0 & 0 & 0 & 0 & 0 & 0 & 0 & 0 & 0 & 0 & 0 \\
0 & 0 & 0 & 0 & 0 & 0 & 0 & 0 & 0 & 0 & 0 \\
0 & 0 & 0 & 0 & 0 & 0 & 0 & 0 & 0 & 0 & 0 \\
0 & 0 & 0 & 0 & 0 & 0 & 0 & 0 & 0 & 0 & 0 \\
0 & 0 & 0 & 0 & 0 & 0 & 0 & 0 & 0 & 0 & 0 \\
0 & 0 & 0 & 0 & 0 & 0 & 0 & 0 & 0 & 0 & 0 \\
0 & 0 & 0 & 0 & 0 & 0 & 0 & 0 & 0 & 0 & 0 \\
0 & 0 & 0 & 0 & 0 & 0 & 0 & 0 & 0 & 0 & 0 \\
0 & 0 & 0 & 0 & 0 & 0 & 0 & 0 & 0 & 0 & 0 \\
0 & 0 & 0 & 0 & 0 & 0 & 0 & 0 & 0 & 0 & 0 \\
\end{bmatrix}
\begin{matrix}
10 \\ 0 \\ 0 \\ 0 \\ 0 \\ 0 \\ 0 \\ 0 \\ 0 \\ 0 \\ 0
\end{matrix}
$$

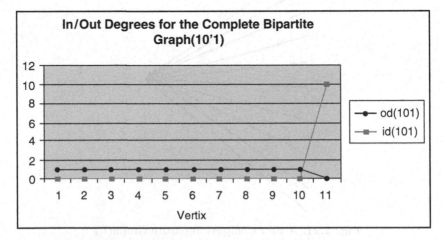

Fig. 2.22. In/Out for CBG(10×1) of figure 2.21

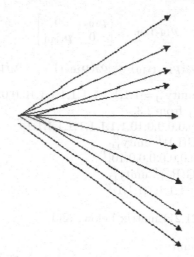

Fig. 2.23. A 1*10 Complete Bipartite Graph(R)

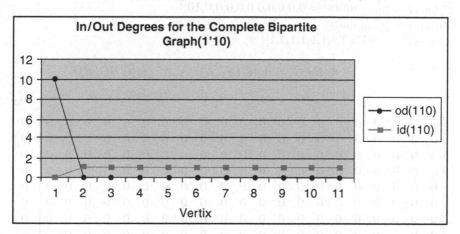

Fig. 2.24. In/Out for CBG(1×10) of figure 2.23

The CBG(R) looks like (See Figure 2.23) and its In/Out degree characteristic in Figure 2.24. We can conclude in general that:

$$A[BTG] = \begin{bmatrix} A(CBG(L)) & 0 \\ 0 & A(CBG(R)) \end{bmatrix}$$

The o_d of a BTG is related to the o_d of the CBG(L) and the o_d of CBG(R):
$o_d(\text{BTG}) = [o_d(\text{CBG(L)} \quad o_d(\text{CBG(R)})]^T$
$o_d(\text{BTG})=A(\text{BTG}).\text{unity}_{21}^T=[1,1,1,1,1,1,1,1,1,1,10,0,0,0,0,0,0,0,0,0,0]^T$
where $unity_{21}$ is the vector of 21 ones. Applying it to the BTG(10*1*10), we conclude that:

$$p_{10*1*10} = \begin{bmatrix} p_{10*1} & 0 \\ 0 & p_{1*10} \end{bmatrix}$$

and: $\qquad\qquad\qquad o_d(p_{10*1*10}) = [o_d(\mathrm{p}_{10*1}) \qquad o_d(\mathrm{p}_{1*10})]^T$

$o_{d(p10*1*10)} = p_{10*1*10}.\mathrm{unity}_{21}^T = [1,1,1,1,1,1,1,1,1,1,10,0,0,0,0,0,0,0,0,0,0]^T$
$i_d(\mathrm{BTG})^T = A^T.\mathrm{unity}_{21}^T$ from 2.3.
$i_d(\mathrm{BTG})^T = [0,0,0,0,0,0,0,0,0,0,0,10,1,1,1,1,1,1,1,1,1,1]^T$
$i_d(\mathrm{CBG(L)})^T = A(\mathrm{CBG(L)})^T.\mathrm{unity}_{11}^T$
$i_d(\mathrm{CBG(L)})^T = [0,0,0,0,0,0,0,0,0,0,10]^T$
$i_d(\mathrm{CBG(R)})^T = A(\mathrm{CBG(R)})^T.\mathrm{unity}_{11}^T$
$i_d(\mathrm{CBG(R)})^T = [0,1,1,1,1,1,1,1,1,1,1]^T$

where $p_{10*1*10}$ is the (21,21) matrix below, and

$i_d(p_{10*1*10}) = \mathrm{p}_{10*1*10}^T*\mathrm{unity}_{21}^T$
$i_d(p_{10*1*10}) = [0,0,0,0,0,0,0,0,0,0,0,10,1,1,1,1,1,1,1,1,1,1]^T$
$i_d(p_{10*1}) = \mathrm{p}_{10*1}^T.\mathrm{unity}^T = [0,0,0,0,0,0,0,0,0,0,10]^T$
$i_d(p_{1*10}) = \mathrm{p}_{1*10}^T.\mathrm{unity}^T$
$i_d(p_{1*10})^T = [0,1,1,1,1,1,1,1,1,1,1]^T$

$$i_d[BTG] = [i_d(CBG(L)) \qquad i_d(CBG(R))]^T$$

$$
\begin{array}{c}
 0_d \\
\left[\begin{array}{ccccccccccccccccccccc}
0&0&0&0&0&0&0&0&0&0&0&0&0&0&0&0&0&0&0&0&0 \\
0&0&0&0&0&0&0&0&0&0&0&0&0&0&0&0&0&0&0&0&0 \\
0&0&0&0&0&0&0&0&0&0&0&0&0&0&0&0&0&0&0&0&0 \\
0&0&0&0&0&0&0&0&0&0&0&0&0&0&0&0&0&0&0&0&0 \\
0&0&0&0&0&0&0&0&0&0&0&0&0&0&0&0&0&0&0&0&0 \\
0&0&0&0&0&0&0&0&0&0&0&0&0&0&0&0&0&0&0&0&0 \\
0&0&0&0&0&0&0&0&0&0&0&0&0&0&0&0&0&0&0&0&0 \\
0&0&0&0&0&0&0&0&0&0&0&0&0&0&0&0&0&0&0&0&0 \\
0&0&0&0&0&0&0&0&0&0&0&0&0&0&0&0&0&0&0&0&0 \\
0&0&0&0&0&0&0&0&0&0&0&0&0&0&0&0&0&0&0&0&0 \\
1&1&1&1&1&1&1&1&1&1&0&0&0&0&0&0&0&0&0&0&0 \\
0&0&0&0&0&0&0&0&0&0&1&0&0&0&0&0&0&0&0&0&0 \\
0&0&0&0&0&0&0&0&0&0&1&0&0&0&0&0&0&0&0&0&0 \\
0&0&0&0&0&0&0&0&0&0&1&0&0&0&0&0&0&0&0&0&0 \\
0&0&0&0&0&0&0&0&0&0&1&0&0&0&0&0&0&0&0&0&0 \\
0&0&0&0&0&0&0&0&0&0&1&0&0&0&0&0&0&0&0&0&0 \\
0&0&0&0&0&0&0&0&0&0&1&0&0&0&0&0&0&0&0&0&0 \\
0&0&0&0&0&0&0&0&0&0&1&0&0&0&0&0&0&0&0&0&0 \\
0&0&0&0&0&0&0&0&0&0&1&0&0&0&0&0&0&0&0&0&0 \\
0&0&0&0&0&0&0&0&0&0&1&0&0&0&0&0&0&0&0&0&0 \\
0&0&0&0&0&0&0&0&0&0&1&0&0&0&0&0&0&0&0&0&0
\end{array}\right]
\begin{array}{c}
0 \\ 0 \\ 0 \\ 0 \\ 0 \\ 0 \\ 0 \\ 0 \\ 0 \\ 0 \\ 10 \\ 1 \\ 1 \\ 1 \\ 1 \\ 1 \\ 1 \\ 1 \\ 1 \\ 1 \\ 1
\end{array}
\end{array}
$$

2.4 Final Observations on Web Graphs

Would the classification of these different Web graphs developed in equation 2.11 help shed a light on the behavior of Kleinberg's algorithm ([64]) even though Kleinberg's algorithm is not related to the (Out-degree, In-degree) space in which all these different graphs have been studied. According to equation 2.11 complete bipartite graphs(CBG) have a behavior close to Out-degree graphs(OD), and that OD graphs have a behavior close to general graphs(GG), and that GG have a behavior close to in-degree graphs (ID). Chapter 3 will reveal whether the general classification scheme proposed was able to help predict the behavior of these graphs. This chapter focuses on discovering important topological structures in Web pages and to help predict the behavior of Web algorithms (see chapter 3) in such environment especially that the WWW is rich not only in information but in topological structures. Last of all, what if a user's Web page did not fit in any of the different Web graphs already described? Can we evaluate such an activity in a given Web page? May be different types of graphs are needed for that? Further research needs to answers the following questions:

1. How do we categorize existing simple graphs such the one already in use in many areas of research?
2. How do we uncover Web algorithms that are efficient on such graphs?
3. How do we devise new graphs, to better characterize user creativity in a given Web page on the WWW?
4. How do we devise new algorithms for these graphs?

The WWW is full of information and mining the Web for different kinds of information is still in its infancy. The difficulty in evaluating information on the Web has to do with our cognitive abilities combined with personal likings, which is hard to quantify always. We are on the verge of unveiling creativity as a graph and what best describes the birth and death of an idea which happens in the birth and death of Web pages and the changes of their content.

3

Link Analysis of Web Graphs

3.1 Introduction

Scientific citations have been studied for a long time. Citation analysis in the field of bibliometrics (*Egghe & Rousseau* [32]) is the science of studying citations, their structure, and the evolution of a specific domain of knowledge from its citations. Many information sciences journals, e.g., JASIS (*Small* [122, 123]) have devoted issues and complete volumes to the exploration of such a field of knowledge. A citation is static and unidirectional. Once an article is published, no new references can be added to it. A citation can be used to evaluate its impact and influence over the whole body of knowledge in a given field. A citation fulfilling such a role in a given field becomes an authority in that field. Citations in a given journal follow the same principles, standards, and forms. Thus, the standard deviation of the content of two articles in a given journal is usually small. Human judgment of the technicality of an article keeps the quality of publication at high level although the noise is present, and thus a citation is more relevant and more objective to a given topic. Citations link articles that are relevant to a given research. Garfield's impact factor ([38]) is the most important factor ever developed to assess a journal j influence over the publication in that field. The impact factor is the average number of citations a journal j in a given year receives from other journal articles after it has been published for the last 2 years. It becomes the in-degrees of nodes in a given network of publications in that field. *Pinski & Narin* ([111]) argued that a journal is influential if it is heavily cited by other influential journals. Citations of a given publication are "signatures of intelligence" of the space and time of a collective group of people.

Hubs and authorities (*Meghabghab & Kandel* [87]) have been developed to better assess the influence that a Web page has on other Web pages that link to it. Web pages also reflect the "signatures of intelligence" in our era and contain rich information on our collective society as a whole. Citation analysis is a very important field for information scientists for mining citations in journals. The comparison above sheds some light on the parallel between a

G. Meghabghab and A. Kandel: *Search Engines, Link Analysis, and User's Web Behavior*, Studies in Computational Intelligence (SCI) **99**, 47–68 (2008)
www.springerlink.com   © Springer-Verlag Berlin Heidelberg 2008

Web page and a citation in a journal ([87]). This just to emphasize the interest that Web mining experts ought to have to analyze Web links even though it is a task in its infancy, and even though that the principles that govern Web pages and Web links are different from those of scientific citations. Web pages are not scrutinized like scientific journals are. New links in a Web page can have an impact on the whole search done and the always changing size of the Web can reflect patterns of indexing, crawling for search engine designers that feed on these Web links. To fully comprehend and assess the discovery of hubs and authorities and the influence that hyperlinks between different Web pages have over other Web pages, graph theory can be used to better understand the measures developed by [64]. The same idea can be used to measure the distance between 2 Web pages. Measuring the topological structure richness of a collection of Web pages is an important aspect of Web pages never explored before and never applied before on hubs and authorities.

3.2 Linear Algebra Applied to Link Analysis

From linear algebra (*Golub & Van Loan* [41]): if A is a symmetric n x n matrix, then an eigenvalue of A is a number g with the property for some vector v we have $Av = gv$. The set of all such v is a subspace of Rn, which we refer to as the eigenspace associated with g; the dimension of this space is the multiplicity of g. A has n distinct eigenvalues, each of them is a real number, and the sum of their multiplicity is n. These eigenvalues are denoted by $g_1(A), g_2(A), .., g_n(A)$ indexed in order of decreasing absolute value and with each value listed in a number of times equal to its multiplicity. For each distinct eigenvalue, we choose an orthonormal basis of its eigenspace. Considering the vectors in all these bases, we obtain a set of eigenvectors $v_1(A), v_2(A), v_3(A), ...v_n(A)$ where $v_i(A)$ belongs to the eigenspace of $g_i(A)$. The following assumption can be made without any violation of any substantial principle in what follows that:

$$g_1(A) > g_2(A) \tag{3.1}$$

When the last assumption holds 3.1, $v_1(A)$ is referred to as the principal eigenvector and all other $v_i(A)$ are non-principal eigenvectors. By applying this analysis to II^T we can conclude:

1. There are 4 distinct eigenvalues mainly: 0, 1, 2, 3 with the multiplicity of g = 1 being 3.
2. The eigenspace can be defined by:

$$3(II^T) > 2(II^T) > 1(II^T) > = 1(II^T) > = 1(II^T) > 0(II^T)$$

3. There is one principal eigenvector which corresponds to the eigenvalue of g=3. This vector has a value $v_3 = [0, 0, 0, 1, 0, 0]^T$. All other eigenvectors are not principal eigenvectors.

3.3 Kleinberg's Algorithm of Link Analysis

In [64], Kleinberg built a search technique procedure which he called HITS, which stands for "Hyperlink Induced Topic Search". It was meant to rank the result set of links by any commercial search engine on broad search topics. The first step in HITS is a sampling step. The result of such a sampling is called the root set. Each page tends to be considered a hub or an authority or neither. HITS try to find "hubs" Web pages and "authorities" Web pages. An authority page is a page that has the highest number of links pointing to it in a given topology. A hub is a Web page that points to a high number of authority Web pages. Authorities and hubs reinforce each other mutually: since Web pages tend to be part of what makes an authority or so close to an authority that they are hubs. There are pages that are neither authorities nor hubs. Good authority Web pages are found close to good hubs, which in turn are pointing towards good sources of information. A typical representation of hubs and authorities can be seen as a bipartite graph. Figure 3.1 represents an ideal situation of hubs and authorities. The root set mentioned above tends to contain few hundred Web pages. To the root set is added links that point to it. This new set can grow to form up to few thousands of Web pages and is called the base set. The second step in HITS is to compute the weight for each Web page that can help rank the links as their relevance to the original query. Even though that HITS is applied to a large number of Web pages not

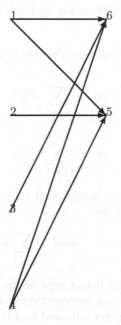

Fig. 3.1. A bipartite graph

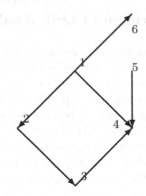

Fig. 3.2. A bipartite graph G_1

individual Web pages, for illustration only, we can look at the Web pages in Figure 3.2. From the actual definition of authority Web pages and hub Web pages we can conclude:

1. Web page 4 is an "authority" Web page because it has the highest number of pages that point to it. Web page 1 is also an authority Web page.
2. Web pages 5 and 3 are good "hub" pages. Web pages 5 and 3 cannot be considered authorities.
3. Web pages 2 and 6 are neither hubs nor authorities.
4. Thus, a Web page in HITS tends to have a hub weight and authority weight.

HITS assigns an authority weight a_p and a hub weight h_p for every Web page that reinforce each other:

$$a_p = \sum_{(q,p)\in R} (h_q) \tag{3.2}$$

$$h_p = \sum_{(q,p)\in R} (a_q) \tag{3.3}$$

These values of a_p and h_p are maintained normalized on all the Web pages for that particular graph. Thus:

$$\sum_{(p)}(a_p^2) = 1 \quad and \quad \sum_{(p)}(h_p^2) = 1 \tag{3.4}$$

The larger the value of a_p the better representative of an authority is p. The higher the value of h_p the better representative of a hub is p. These values get initialized first to 1 and then get adjusted and the first few hubs and the first few authorities are then filtered out on a given graph. The procedure Filter looks like the following: Filter F:

1. Let z is a vector $z = [1,1,1,..1]^T$ of R_n where n is the number of Web pages considered.
2. Initialize a_{p0} and h_{p0} to z
3. For $i = 1, 2, ..k$

   (a) Use 3.2 to (a_{pi-1}, h_{pi-1}) obtaining new a weights a_i'
   (b) Use 3.3 to (a_{pi}', h_{pi-1}) obtaining new h weights h_i'
   (c) Normalize a_{pi}', and get a p_i
   (d) Normalize h_i', and get h_i
   (e) End
   (f) Return (a_{pk}, h_{pk})

We are proposing the following three conjectures:

1. conjecture 1: It turns out that the returned set (a_{pk}, h_{pk}) is made out of 2 values where a_{pk} is the principal eigenvector over $A^T A$, and h_{pk} is the eigenvector over AA^T.
2. conjecture 2: It turns out that that a Web page cannot be an authority Web page and a hub Web page after the first iteration according to Kleinberg's Filter Procedure.
3. conjecture 3: It turns out that that a Web page cannot be an authority Web page and a hub Web page at the end of procedure Filter according to Kleinberg's Filter Procedure.

3.4 Applying HITS to the Graphs of 2.2

3.4.1 HITS Applied to General Graphs

Let us apply the procedure Filter F to Figure 3.2. The number of Web pages in Figure 3.2 is n = 6. Parameters a and h after one single iteration of step 3 in F and before any normalization have the following values: a = $[2,1,1,3,0,1]^T$ and h = $[5,1,5,0,5,0]^T$. After normalization:
a = $[.5,.25,.25,.75,0,.25]^T$ and h = $[.5735,.1147,.5735,0,.5735,0]^T$.
Here are some of the observations after the first iteration:

1. Web page 4 is the best authority Web page.
2. Web pages 1, 3 and 5 are the best hub pages.
3. Web page 2 is the second best hub Web page.
4. Web pages 2, 3, and 6 are the 3rd best authority Web pages.
5. Web pages 4 and 6 are not hub Web pages.
6. Web Pages 1, 2, and 3 are authority and hub Web pages at the same time with different ranks.

General Graphs make appear the idea of a Web page that is both an authority Web page and a hub page in the first iteration. Web pages 1,2, and 3 in Figure 3.2 prove that to a certain extent. Would that idea persist after a number of iterations? conjecture 2 is not verified in the case of General

Graphs. What happens after a number of iterations? The same idea we used in the subsection on the "distance between Web pages" can be applied here. Iterate until the values of vectors a and h do not change any more. What happened to conjecture 3 in the case of General Graphs? Here are the final values of a and h after k = 6 iterations with a change of almost 0: $a = [.5164,.2582,0,.7746,0,.2582]^T$, $h = [.5774,0,.5774,0,.5774,0]^T$. Here are some final observations after k=6 iterations in the procedure Filter:

1. Web Page 4 is still the best authority Web page.
2. Web pages 1, 3 and 5 are still the best and only hub Web pages.
3. Web page 1 is now the 2nd best authority Web page.
4. Web pages 2 and 6 are now the 3rd best authority Web pages.
5. Web page 3 is not an authority Web page any more.
6. Web page 1 is now the only hub Web page and authority Web page.

Web page 1 is still the key issue to tackle in the whole subject of hubs and authorities. It was not well expounded in the literature whether that is a major obstacle to the full expansion of hubs and authorities in general. conjecture 3 does not hold for General Graphs. Being satisfied with the early set of iterations for a and h would have been deceiving. Some more values which had a non zero value now have a zero value. The answer to that question lies in the fact that we want to verify that the final values of vectors a and h correspond to the eigenvectors of both $A^T A$ and AA^T correspondingly. From $D = A^T A$ in section I. we can calculate the eigenvalues and eigenvectors of D. The columns of the following matrix M_1 constitute the eigenvectors:

$$
M_1 = \begin{bmatrix}
0 & -.1647 & -.607 & -.0665 & .5774 & .5164 \\
0 & -.2578 & -.182 & -.7074 & -.5774 & .2582 \\
-1 & 0 & -0 & -0 & -0 & -0 \\
-0 & .1647 & .607 & .0665 & -0 & .7746 \\
-0 & .9331 & -.2221 & -.2829 & 0 & 0 \\
-0 & .0931 & -.425 & .6409 & -.5774 & .2582
\end{bmatrix}
$$

And the corresponding eigenvalue vector:
eigenvector(D)=$(1.0,0.0,0.0,0.0,2.0,5.0)^T$
According to our notations in section 2 we have:

$$5(A^T A) > 2(A^T A) > 1(A^T A) > 0(A^T A) \geq 0(A^T A) \geq 0(A^T A)$$

So the eigenvalue is 5.0 which is the last value of the eigenvector(D). The corresponding eigenvector is a vector denoted by $w_6(M_1)$ which the last column of M_1: $w_6(M_1) = [0.5164, 0.2582, -0.0, 0.7746, 0.0, 0.2582]^T$ By comparing the last vector and the vector a already calculated above we conclude:

$$w_6(M_1) = a \tag{3.5}$$

From $C = AA^T$ in section 2, we can calculate the eigenvalues and eigenvectors of C. The columns of the following matrix M_2 constitute the eigenvectors:

$$M_2 = \begin{bmatrix} 0 & 0 & 0 & .8165 & .5774 & 0 \\ 1 & 0 & 0 & 0 & 0 & 0 \\ 0 & .2571 & -.6587 & -.4082 & .5774 & 0 \\ 0 & .9316 & .3635 & -0 & 0 & 0 \\ 0 & -.2571 & .6587 & -.4082 & 5774 & 0 \\ 0 & 0 & 0 & 0 & 0 & 1 \end{bmatrix}$$

Here is the corresponding eigenvalue vector:$[1.0, 0.0, 0.0, 2.0, 5.0, 0.0]^T$ So according to our notations in section 2, we have:

$$5(AA^T) > 2(AA^T) > 1(AA^T) > 0(AA^T) \geq 0(AA^T) \geq 0(AA^T)$$

So the eigenvalue is 5.0 which is the 5th value in the vector of eigenvalue. So the principal eigenvector denoted by $w_5(M_2)$ which the fifth column of D would be:$[0.5774,0,0.5774,0,0.5774,0]^T$ By comparing the last vector and the vector h already calculated above we conclude:

$$w_5(M_2) = h \tag{3.6}$$

3.4.2 HITS applied to In-degree Graphs

Let us apply the Filter procedure to an In-degree Graph like the one in Figure 3.3.

The number of Web pages in Figure 3.3 is n = 8. Parameters a and h after one single iteration of step 3 in F and before any normalization have the following values: a $= [0,0,0,0,0,2,2,3]^T$ and h $= [2,2,3,2,2,3,3,0]^T$. After normalization a $= [0,0,0,0,0.4851,0.4851,0.7276]^T$ and h $= [0.305,0.305,0.4575, 0.305,0.305, 0.4575,0.4575,0]^T$. Here some of the observations that can be made from a and h after one single iteration (k = 1):

1. Web Page 8 is the best authority Web page.
2. Web pages 6, and 7 are the 2nd best authority Web pages.

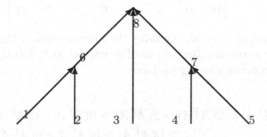

Fig. 3.3. An In Degree Graph

3. Web pages 3, 6 and 7 are the best hub Web pages.
4. Web pages 1, 2, 4, and 5 are the 2nd best hub Web pages.
5. Web pages 6 and 7 are authority Web pages and hub Web pages at the same time.

The first iteration of the procedure Filter shows a Web page that is both an authority Web page and a hub page. Web pages 6 and 7 in that example show clearly that conjecture 2 does not hold for In-degree Graphs. Would that idea persist after a number of iterations? What happened to conjecture 3 in the case of In-degree Graphs. Here are the final values of a and h after k = 8 iterations with a change of almost 0: a = $[0, 0, 0, 0, 0, 0.039, 0.039, 0.9985]^T$ h = $[0.02, 0.02, 0.5768, 0.02, 0.02, 0.5768, 0.5768, 0]^T$. Here are the final observations after k = 8 iterations on the procedure Filter:

1. Web Page 8 is the only authority Web page.
2. Web pages 3,6, and 7 are the only hub pages.

The process of identifying hubs and authorities is an iterative process and the first iteration is just the beginning of the process of filtering out weak hubs and authorities. In-Degree trees seem to be ideal for the procedure Filter since no one Web page can be considered both a hub and an authority at the same time. conjecture 3 holds for In-degree Graphs. Being satisfied with the early set of iterations for a and h would have been deceiving. Some more values which had a non zero value now have a zero value. The answer to that question lies in the fact that we want to verify that the final values of vectors a and h correspond to the eigenvectors of both $A^T A$ and AA^T correspondingly. We want still to verify conjecture 1. It turns out that $D = A^T A$ is a diagonal matrix. The eigenvalues and eigenvectors can be calculated simply with no other transformations like what we did in the case of General Graphs.

$$M_1 = \begin{bmatrix} 0 & 0 & 0 & 0 & 0 & 0 & 0 & 0 \\ 0 & 0 & 0 & 0 & 0 & 0 & 0 & 0 \\ 0 & 0 & 0 & 0 & 0 & 0 & 0 & 0 \\ 0 & 0 & 0 & 0 & 0 & 0 & 0 & 0 \\ 0 & 0 & 0 & 0 & 0 & 0 & 0 & 0 \\ 0 & 0 & 0 & 0 & 0 & 2 & 0 & 0 \\ 0 & 0 & 0 & 0 & 0 & 0 & 2 & 0 \\ 0 & 0 & 0 & 0 & 0 & 0 & 0 & 3 \end{bmatrix}$$

The corresponding matrix M_1 contains the eigenvectors is the unit matrix of 8*8. Here is the corresponding eigenvalue vector: $[0, 0, 0, 0, 0, 2, 2, 3]^T$ So according to our notations in 3.2 we have:

$$3(A^T A) > 2(A^T A) > 2(A^T A) > 0(A^T A > 0(A^T A)$$
$$\geq 0(A^T A) \geq 0(A^T A \geq 0(A^T A).$$

The eigenvalue is 5.0 which is the 8th value in the vector of eigenvalue. So the principal eigenvector denoted by $w_8(M_1)$ which the last column of M_1 would be:$[0,0,0,0,0,0,0,1]^T$. By comparing the last vector and the vector a already calculated above we conclude:

$$w_8(M_1) = a \qquad (3.7)$$

From $C = AA^T$, we can calculate the eigenvalues and eigenvectors of C. The columns of the following matrix M_2 constitute the eigenvectors:

$$M_2 = \begin{bmatrix} .707 & .707 & 0 & 0 & 0 & 0 & 0 & 0 \\ -.707 & .707 & 0 & 0 & 0 & 0 & 0 & 0 \\ 0 & 0 & 0 & 0 & .785 & .225 & .578 & 0 \\ 0 & 0 & -.707 & .707 & 0 & 0 & 0 & 0 \\ 0 & 0 & .707 & .707 & 0 & 0 & 0 & 0 \\ 0 & 0 & 0 & 0 & -.587 & .568 & .578 & 0 \\ 0 & 0 & 0 & 0 & -.198 & -.792 & .578 & 0 \\ 0 & 0 & 0 & 0 & 0 & 0 & 0 & 1 \end{bmatrix}$$

The corresponding eigenvalue vector $= [0,2,0,2,0,0,3,0]^T$ According to our notations in 2, we have:

$$3(AA^T) > 2(AA^T) > 2(AA^T) > 0(AA^T)$$
$$\geq 0(AA^T) \geq 0(AA^T) \geq 0(AA^T) \geq 0(AA^T)$$

The eigenvalue is 3 which is the 7th value in the vector of eigenvalue. The principal eigenvector denoted by $w_7(M_2) = [0,0,0.5774,0,0,0.5774,0.5774,0]^T$ By comparing the last vector and the vector h already calculated above we conclude:

$$w_7(M_2) = h \qquad (3.8)$$

3.4.3 HITS applied to Out-degree Graphs

Let us apply the Filter procedure to an out-degree Graph like the one in Figure 3.4.

The number of Web pages in Figure 3.4 is n = 8. Parameters a and h after one single iteration of step 3 in F and before any normalization have the following values: $a = [0,1,1,1,1,1,1,1]^T$ and $h = [2,3,2,0,0,0,0,0]^T$. The values of a and h after normalization are: $a = [0,.378,.378,.378,.378,.378,0,.378,.378]^T$ and $h = [.4581,.7276,.4581,0,0,0,0,0]^T$. Here are some observations that can be made from the first iterative values of a and h:

1. Web Pages 2,3,4,5,6,7, and 8 are authority pages.
2. Web page 2 is the best hub Web page.
3. Web pages 1and 3 are the 2nd best hub Web pages.

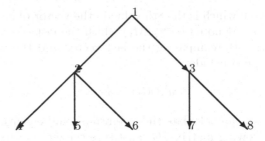

Fig. 3.4. An Out Degree Graph

4. Web pages 2 and 3 are hub Web pages and authority Web pages at the same time.

The first iteration of the procedure Filter shows a Web page that is both an authority Web page and a hub page. Web pages 2 and 3 in that example show clearly that. conjecture 2 does not hold for Out-degree graphs. Would that idea persist after a number of iterations. What happened to conjecture 3 in the case of out-degree graphs. Here are the final values of a and h after k = 8 iterations with a change of almost 0:
$a = [.03, .03, .03, .576, .576, .576, .03, .03]^T$, $h = [.04, .999, .04, 0, 0, 0, 0, 0]^T$.
Here are some final observations after k = 8 iterations on the procedure Filter:

1. Web Pages 4,5, and 6 are the only authority Web pages.

2. Web page 2 is the only hub Web page.

This final result requires some interpretations. Even though Web page 3 could have been considered a hub Web page but because in the same graph more nodes came out of Web page 2 than Web page 3, the measure of a hub for a Web page is a global measure and not a local one. Second, that because 2 is the nub page, Web pages 4, 5, and 6 become the only authority Web pages and not any more 7 and 8 like it was in the first iteration of the procedure Filter. Again the idea of an authority Web page is a global measure and not a local one. This idea could be applied to Citation analysis in general, which says that in a given literature on a given topic, being an authority is not a second best but really the best. The more you are cited, the more links come to your Web page and the more attention you will receive, and that make you an authority in that field regardless of the second best. Further studies have reflected on the fact that there may be more than just authority in a given graph, which should be considered. conjecture 3 does hold for Out-degree graphs. Being satisfied with the early set of iterations for a and h would have been deceiving. Some more values which had a non zero value now have a zero value. The answer to that question lies in the fact that we want to verify that the final values of vectors a and h correspond to the eigenvectors of both $A^T A$ and AA^T correspondingly. To verify conjecture 1, we calculate $D = A^T A$. we can calculate the eigenvalues and eigenvectors of D. The columns of the following matrix M_1 constitute the eigenvectors:

$$M_1 = \begin{bmatrix} 1 & 0 & 0 & 0 & 0 & 0 & 0 & 0 \\ 0 & .707 & .707 & 0 & 0 & 0 & 0 & 0 \\ 0 & .707 & -.707 & 0 & 0 & 0 & 0 & 0 \\ 0 & 0 & 0 & .785 & .225 & .577 & 0 & 0 \\ 0 & 0 & 0 & -.587 & .568 & .577 & 0 & 0 \\ 0 & 0 & 0 & -.198 & -.792 & .577 & 0 & 0 \\ 0 & 0 & 0 & 0 & 0 & 0 & .707 & .707 \\ 0 & 0 & 0 & 0 & 0 & 0 & .707 & -.707 \end{bmatrix}$$

Here is the corresponding eigenvalue vector: $[0,2,0,0,0,3,2,0]^T$. Acording to our notations in 3.2 we conclude that:

$$3(AA^T) > 2(AA^T) > 2(AA^T) > 0(AA^T) >$$
$$0(AA^T) \geq 0(AA^T) \geq 0(AA^T) \geq 0(AA^T)$$

The eigenvalue is 3 which is the 6th value in the vector of eigenvalue. The principal eigenvector denoted by $w_6(M_1)$ would be the principal eigenvector. The principal eigenvector denoted by $w_6(M_1)=[0,0,0,.5774,.5774,.5774,0,0]^T$ By comparing the last vector and the vector a already calculated above we conclude:

$$w_6(M_1) = a \qquad (3.9)$$

It turns out that $C = AA^T$ is a diagonal matrix. The eigenvalues and eigenvectors can be calculated simply with no other transformations like what we did in the case of General Graphs.

$$M_1 = \begin{bmatrix} 2 & 0 & 0 & 0 & 0 & 0 & 0 & 0 \\ 0 & 3 & 0 & 0 & 0 & 0 & 0 & 0 \\ 0 & 0 & 2 & 0 & 0 & 0 & 0 & 0 \\ 0 & 0 & 0 & 0 & 0 & 0 & 0 & 0 \\ 0 & 0 & 0 & 0 & 0 & 0 & 0 & 0 \\ 0 & 0 & 0 & 0 & 0 & 0 & 0 & 0 \\ 0 & 0 & 0 & 0 & 0 & 0 & 0 & 0 \\ 0 & 0 & 0 & 0 & 0 & 0 & 0 & 0 \end{bmatrix}$$

The corresponding matrix M_2 contains the eigenvectors is the unit matrix of 8*8. Here is the corresponding eigenvalue vector:$[2,3,2,0,0,0,0,0]^T$ Thus according to our notations in section 3.2, we conclude that:

$$3(A^T A) > 2(A^T A) > 2(A^T A) > 0(A^T A)$$
$$> 0(A^T A) = 0(A^T A) = 0(A^T A) = 0(A^T A).$$

The eigenvalue is 3 which is the 2th value in the vector of eigenvalue. The principal eigenvector denoted by $w_2(M_2)$ would be the principal eigenvector. The principal eigenvector denoted by $w_2(M_2)=[0,1,0,0,0,0,0]^T$ By comparing the last vector and the vector h already calculated above we conclude:

$$w_2(M_2) = h \qquad (3.10)$$

3.4.4 HITS Applied to "Complete Bipartite" Graphs

Let us apply the Filter procedure to a "complete bipartite" Graph like the one in Figure 3.5. The number of Web pages in Figure 3.5 is n=7. Parameters a and h after one single iteration of step 3 in F and before any normalization have the following values: a=$[0,0,0,0,4,4,4]^T$ and h=$[12,12,12,12,0,0,0]^T$. After normalization:

a=$[0,0,0,0,0.5774,0.5774,0.5774]^T$ and h=$[0.5,0.5,0.5,0.5,0,0,0]^T$.

The following observations can be made from a and h after the first iteration:

1. Web Pages 5, 6, and 7 are the only authority pages.
2. Web pages 1,2,3 and 4 are the only hub pages.

In this perfect case, no pages are both authority and hub pages at the same time. In this case also, there are no Web pages that are neither authority Web pages nor hub Web pages. The observations made in the first iteration are evident from the complete bipartite graph itself. conjecture 2 holds for complete bipartite graphs. What happened to conjecture 3 in the case of complete bipartite graphs? Complete bipartite graphs are ideal cases for finding hubs and authorities. We need not iterate any further the procedure Filter since

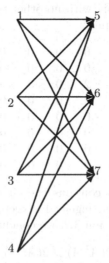

Fig. 3.5. A complete bipartite graph

no change will occur any way. But still one can iterate to verify that the values of a and h will stay the same. Here are the final values of a and h after k=8 iterations with no change: a=$[0,0,0,0,0.5774,0.5774,0.5774]^T$ and h=$[0.5,0.5,0.5,0.5,0,0,0]^T$. The early observations made on a and h will stand after a number of iterations. Conjecture 3 holds for complete bipartite graphs.

3.4.5 HITS Applied to Bipartite Graphs

Let us apply the Filter procedure to a bipartite graph like the one in Figure 3.6. The number of Web pages in Figure 3.6 is n=6. Parameters a and h after one single iteration of step 3 in F and before any normalization have the following values: a=$[0,2,0,2,0,1]^T$ and h=$[3,0,4,0,2,0]^T$. After normalization the values of a and h are: a=$[0,0.6667,0,0.6667,0,0.3333]^T$ and h=$[0.5571,0,0.7428,0,0.3714,0]^T$. The following observations can be made from the first values of a and h:

1. Web Page 2 and 4 are the best authority pages.
2. Web page 3 is the best hub page.
3. Web page 6 is the 2nd best authority Web page.
4. Web page 1 is the 2nd best hub Web page.
5. Web page 5 is the 3rd best hub Web page.
6. No Web pages are authority Web pages and hub pages at the same time.
7. All Web pages are either authority Web pages or hub Web pages.

conjecture 2 holds for bipartite graphs. What happened to conjecture 3 for bipartite graphs? Here are the final values of a and h after k=12 iterations with almost no change: a=$[0,.737,0,.591,0,.328]^T$, h=$[.591,0,.7374,0,.328,0]^T$. The final observations can be made from the final values of a and h:

1. Web Page 2 is the only best authority Web page.
2. Web page 3 is still the best hub Web page.

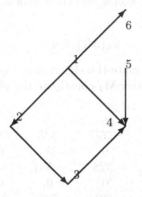

Fig. 3.6. A bipartite graph

3. Web page 4 becomes the 2nd best authority Web page
4. Web page 6 is the 3rd best authority Web page.
5. Web page 1 is still the 2nd best hub Web page.
6. Web page 5 is the still 3rd best hub Web page.
7. No Web pages are authority Web pages and hub pages at the same time.
8. All Web pages are either authority Web pages or hub Web pages.

conjecture 3 holds for bipartite graphs. Being satisfied with the early set of iterations for a and h would have been deceiving. Some more values which had a non zero value now have a zero value. The answer to that question lies in the fact that we want to verify that the final values of vectors a and h correspond to the eigenvectors of both $A^T A$ and AA^T correspondingly. To verify conjecture 1, we calculate $D = A^T A$. Also we need to calculate the eigenvalues and eigenvectors of D. The columns of the following matrix M_1 constitute the eigenvectors:

$$
M_1 = \begin{bmatrix}
1.0 & 0 & 0 & 0 & 0 & 0 \\
0 & 0 & 0 & -.737 & -.328 & .591 \\
0 & 0 & 1.0 & 0 & 0 & 0 \\
0 & 0 & 0 & -.591 & .737 & -.328 \\
0 & 1.0 & 0 & 0 & 0 & 0 \\
0 & 0 & 0 & -.328 & -.591 & -.737
\end{bmatrix}
$$

Here is the corresponding eigenvalue vector: $[0,0,0,3.247,1.555,0.1981]^T$. Thus according to our notations in section 2.3, we have:

$$3.247(AA^T) > 1.555(AA^T) > 0.1981(AA^T) > 0(AA^T) \geq 0(AA^T) \geq 0(AA^T)$$

Thus 3.247 is the eigenvalue which is the 4th value in the vector of eigenvalue. The principal eigenvector would be $w_4(M_1)$. The value of the principal eigenvector denoted by $w_4(M_1)$ would be:$[0,-0.737,0,-0.5910,0,-0.3280]^T$. The reflections made before on using just the value v_i which corresponds to g_i is contested here. We should use the absolute value of $v_i(w)$ as we did with $g_i(A)$. By comparing the last vector and the vector a already calculated above we conclude:

$$w_4(M_1) = a \tag{3.11}$$

From $C = AA^T$, we can calculate the eigenvalues and eigenvectors of C. The columns of the following matrix M_2 constitute the eigenvectors:

$$
M_2 = \begin{bmatrix}
0 & 0 & -.737 & .328 & .591 & 0 \\
0 & 1 & 0 & 0 & 0 & 0 \\
0 & 0 & .328 & -.591 & .737 & 0 \\
1 & 0 & 0 & 0 & 0 & 0 \\
0 & 0 & .591 & .737 & .328 & 0 \\
0 & 0 & 0 & 0 & 0 & 1.0
\end{bmatrix}
$$

Here is the corresponding eigenvalue vector: $[0,0,1.555,0.1981,3.247,0]^T$. Thus according to our notations in 2.3 we have:

$$3.247(AA^T) > 1.555(AA^T) > 0.1981(AA^T) > 0(AA^T) \geq 0(AA^T) \geq 0(AA^T)$$

The eigenvalue is 3.247 which is the 5th value in the vector of eigenvalue. The principal eigenvector would be $w_5(M_2)$. The value of the principal eigenvector denoted by $w_5(M_2)$ would be: $[0.591,0,0.737,0,0.328,0]^T$. By comparing the last vector and the vector h already calculated above we conclude:

$$w_5(M_2) = h \qquad (3.12)$$

3.4.6 Summary of the Application of the HITS Algorithm to the Graphs of 2.2

Did the classification of the web graphs of 2.2 developed in equation 2.11 in section 2.5 of chapter 2 help shed a light on the behavior of Kleinberg's algorithm ([64]) even though Kleinberg's algorithm is not related to the (Out-Degree, In-degree) space in which all these different graphs have been studied. According to equation 2.11 complete bipartite graphs(CBG) have a behavior close to out-degree graphs(OD), and that OD graphs have a behavior close to general graphs(GG), and that GG have a behavior close to in-degree graphs (ID). Conjecture 2 held for CBG, BG but not OD, GG, and ID. Conjecture 2 violated the classification scheme established in that space in the case of OD but held it in the case of GG and ID. Conjecture 3 held for CBG, OD, BG, ID but not GG. According to 2.11 conjecture 3 should not have held ground for ID given the fact that their behavior is close to GG than the other graphs. New classification schemes are needed to help predict the behavior of HITS web algorithms in the new space. The application of HITS algorithm on Bow tie graphs will be considered next.

3.5 Hubs and Authorities in Bow Tie Graphs

In Chapter 2 section 3, we established the relationship between a BTG and the left complete bipartite graph and the right complete bipartite graph. We established that the adjacency matrix of the BTG is made out of the adjacency of the left CBG and the right CBG. The number of inlinks of the BTG is made out of the inlinks of the CBG(L) and the CBG(R). The number of outlinks of the BTG is made out of the outlinks of the CBG(L) and the number of outlinks of the CBG(R). The latter was a special case to begin the study of BTG's. In this section, we start out with a BTG (N*m*N) as a set of 2 complete bipartite graphs CBG with $m << N$, a left one CBG(L) and a right one CBG(R). (They share a core in the middle as seen in figure 2.13). In this ideal case, the BTG behaves very well according to (Meghabghab [95]). And then we consider the challenging cases:

1. A general bow tie graph: BTG($N \times m \times M$) with $m << N$ and $m <<$ M, with a CBG(L) and a CBG(R) but where $m << N$ and $m << M$ and we consider the cases of $N > M$ and $N < M$. In the case of $N <$ M, we discover that all hubs in the left side of the BTG are lost while the authority of the core is lost. In the case of $N > M$, we discover that all authorities in the right side of the BTG are lost while the hub of the core is lost.

2. A general bow tie graph: BTG ($N \times m \times M$) with $m << N$ and $m << M$, as made out of 2 graphs: a complete BG on the left CBG(L) and bipartite graph on the right BG(R) or a BG(L) and CBG(R), which are reasonable cases to study and with $N > M$ or $N < M$. If $N < M$ or $N > M$, and a CBG(L) and a BG(R), we discover that the hub of the core is lost while all the authorities on the right side are lost. If $N < M$ or $N > M$ and a BG(L) and a CBG(R), we discover that the authority of the core is lost while all the hubs on the left are lost.

3. A general bow tie graph: BTG($N \times m \times M$) with $m < N$ and $m < M$, as made out of 2 graphs: a CBG(L) and BG(R) or a BG(L) and CBG(R), which are reasonable cases to study and with $N > M$ or $N < M$. In the case of $N < M$ and a BG(L) and a CBG(R), we discover that the hubs of BG(L) are lost. In the case of a CBG(L) and BG(R) and $N = M$, all hubs of CBG(L) are lost. In the case of a CBG(L) and a BG(R) and $N > M$, then all authorities are lost including the ones of the core. There is less loss of hubs and authorities in this case than in the cases of (1) and (2).

4. Blondel's central score (*Blondel and VanDooren* [9]) was applied to the bow tie graphs of (1), (2), and (3) to help verify whether these cases are different and whether these BTG's are similar. We show that where $m < N$ and $m < M$, the central score(s) identifies that the bow tie graphs of the form N*m*M are similar whether $N > M$ or $N < M$ as long as both are CBG(L) and CBG(R). Also the case of: $N > M$, CBG(L) and BG(R) and the case of: $N < M$, BG(L) and CBG(R), then both BTG are similar.

It turns out that bow tie graphs fall into the category of General Graphs. As seen from table 3.1, in many of the different categories of BTG, the hubs in the left side and the core of the BTG were lost, the authorities of the core and the right side of the BTG were lost. This was never before researched in the study of hubs and authorities in the literature. In the case of table 3.2, there were 2 cases of hubs being lost on the left side of the BTG, and 1 case where the authorities of the core and the right side of the BTG were lost. A study of similarity between these different cases is needed as seen in the next section. Here are the results of hubs and authorities applied to different sizes of bow tie graphs:

Table 3.1. Hubs and authorities in the case of the size of the core $m \ll N$ and $m \ll M$

bow tie graphs $m \ll N, m \ll M$	hubs	authorities
$N*m*N$	Does work	Does Work
$N*m*M(N>M)$	Lost the hubs of the core	Lost the authorities in the BTG(R)
$N*m*M(N<M)$	Lost the hubs in BTG (L)	Lost authorities of the core
$N*m*N$ $N*m$ is a CBG $m*N$ is not CBG	Lost the hubs of the core.	Lost authorities in BTG(R)
$N*m*N$ $N*m$ is not CBG $m*N$ is a CBG	Lost the hubs in BTG(L)	Lost the authorities of the core
$N*m*M$ $N*m$ is a CBG $m*M$ is not CBG and $N>M$	Lost the hubs of the core	Lost the authorities in BTG(R)
$N*m*M$ $N*m is not a CBG$ $m*M$ is a CBG, $N>M$	Lost the hubs in BTG(L)	Lost the authorities of the core
$N*m*M$ $N*m$ is CBG $m*M$ is not CBG, $N<M$	Lost the hubs of the core	Lost the authorities in BTG(R)
$N*m*M$ $N*m$ is not CBG $m*M$ is a CBG, $N<M$	Lost all the hubs in BTG(L)	Lost the authorities of the core

3.6 Similarity of Bow Tie Graphs

Blondel & VanDooren [9] extended the idea of hubs and authority to the general case of a line graph $1 \to 2 \to 3$. The mutually reinforcing updating iteration used above 3.2 and 3.3 can be generalized to the line graph. With each vertex j of G we now associate three scores $x_{i1}, x_{i2}, and x_{i3}$; one for each vertex of the structure graph. We initialize these scores with some positive value and then update them according to the following mutually reinforcing relation. Assume that we have two directed graphs G_A and G_B with n_A and n_B vertices and edge sets E_A and E_B. We think of G_A as a structure graph that plays the role of the graphs hub $\to$ authority and $1 \to 2 \to 3$ in the above examples. We consider real scores x_{ij} for i = $1...n_B$ and j = $1..n_A$ and simultaneously update all scores according to the following updating equations:

$$x_{ij} \leftarrow \sum_{r:(r,i)\in E_B, s:(s,j)\in E_A} (x_{rs}) + \sum_{r:(i,r)\in E_B, s:(j,s)\in E_A} (x_{rs}) \quad (3.13)$$

Table 3.2. Hubs and authorities in the case of the size of the core $m < N$ and $m < M$

bow tie graphs $m < N, m < M$	hubs	authorities
$N * m * M (N > M)$	None were Lost	None were Lost
$N * m * M (N < M)$	None were Lost	None were Lost
$N * m * N (N < M)$	Lost the hubs in BTG (L)	None were Lost
$N * m * N$ $N * m$ is a CBG $m * N$ is not CBG	None were Lost	None were lost
$N * m * N$ $N * m$ is not CBG $m * N$ is a CBG	None were Lost	Lost the authorities of the core and BTG(R)
$N * m * M$ $N * m$ is a CBG $m * M$ is not CBG and $N > M$	None were Lost	None were lost
$N * m * M$ $N * misnotaCBG$ $m * M$ is a CBG, $N > M$	None were Lost in BTG(L)	None were lost
$N * m * M$ $N * m$ is CBG $m * M$ is not CBG, $N < M$	None were Lost	None were lost
$N * m * M$ $N * m$ is not CBG $m * M$ is a CBG, $N < M$	Lost all the hubs in BTG(L)	None were Lost

The updating equation 3.13 can also be written in more compact matrix form. Let X_k be the $n_B * n_A$ matrix of entries x_{ij} at iteration k. Then the updating equations take the simple form:

$$X_{k+1} = B X_k A^T + B^T X_k A; k = 0, 1 \qquad (3.14)$$

Where A and B are the adjacency matrices of G_A and G_B. *Blondel & VanDooren* [9] proved that the normalized even and odd iterates of this updating equation converge, and that the limit of all even iterates is among all possible limits the only one with largest 1-norm. We take this limit as definition of the similarity matrix. A direct algorithmic transcription of 3.14 leads to an approximation algorithm for computing similarity matrices of graphs: Procedure Similarity ([9])

1. Set $Z_0 = 1$.
2. Iterate an even number of times:

   $Z_{k+1} = B Z_k A^T + B^T Z_k A_k / |B Z_k A^T + B^T Z_k A_k|$

   and stop upon convergence of Z_k.
3. Output S.

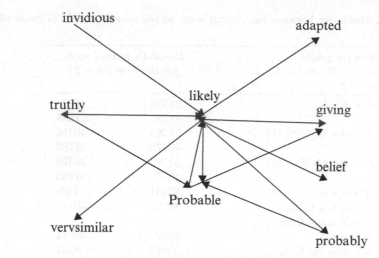

Fig. 3.7. The BTG of "likely" and "probable"

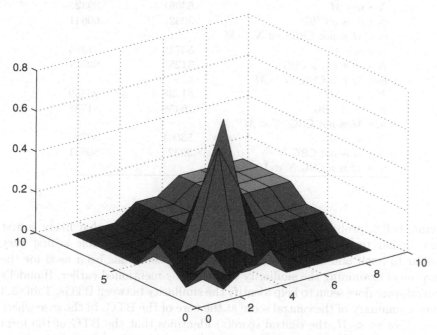

Fig. 3.8. A mesh of the centrality Matrix of Likely and probable

Blondel & VanDooren [9] applied such a similarity to the extraction of synonyms between terms. Figure 3.7 illustrates the graph of "likely" as compared to "probable". Figure 3.8 illustrates a 3-D map of the centrality scores of the nodes 1,2,3,4,5,6,7,8,9 calculated according to the procedure Similarity as they correspond to {invidious, truthy, verisimilar, likely, probable, adapted,

Table 3.3. Similarity between the central score of the core of the BTG made of 2 nodes

bow tie graphs $m << N, m << M$ $m = 2$	Blondel's central score for the core $(m = 2)$	
$N * m * N$	.51538	.4123
	.4123	.51538
$N * m * M(N > M)$	.51783	.40276
	.40276	.51783
$N * m * M(N < M)$	.51783	.40276
	.40276	.51783
$N * m * N$	.53431	.33395
$N * m$ is a CBG	.33395	.6011
$m * N$ is not CBG		
$N * m * N$	.4092	.35074
$N * m$ is not CBG	.35074	.58457
$m * N$ is a CBG		
$N * m * M$	.53061	.3032
$N * m$ is a CBG	.3032	.60641
$m * M$ is not CBG and $N > M$		
$N * m * M$	.54749	.31285
$N * m$ is not a CBG	.31285	.62571
$m * M$ is a CBG, $N > M$		
$N * m * M$	.61738	.26459
$N * m$ is CBG	.26459	.61738
$m * M$ is not CBG, $N < M$		
$N * m * M$	.53061	.3032
$N * m$ is not CBG	.3032	.60641
$m * M$ is a CBG, $N < M$		

giving, belief, probably}. The peak correspond to the fact that likely is first very similar to itself followed by probable, and that probable is first very similar to itself and then to likely. This central score has been used for the purpose of examining the similarity of the BTGs mentioned earlier. Blondel's central score does seem to help identify the similarity between BTGs. Table 3.3 gives a summary of the central score at the core of the BTG. In the case where $m < N$ or $m < M$, the central score(s) identifies that the BTG of the form $(N \times m \times M)$ are similar whether $N > M$ or $N < M$ as long as both are CBG(L) and CBG(R). In the case of $N > M$, CBG(L) and BG(R) and the case of: $N < M$, BG(L) and CBG(R), then both BTG are similar. Figure 3.9 illustrates a 3-D map of the central scores of the $BTG(50 \times 2 \times 100)$ as calculated in the procedure Similarity. The peak in fig. 3.9 corresponds to the nodes 51 and 52, the 2 nodes at the core of the BTG, where 51 is very similar to itself first and then to 52, and 52 is very similar to itself first and then to 51.

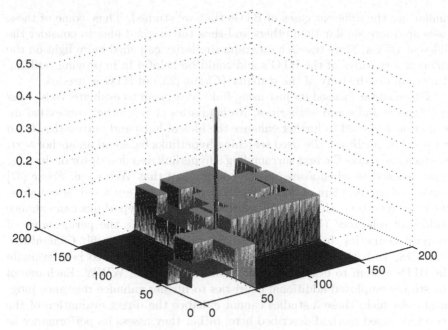

Fig. 3.9. A mesh of the centrality Matrix of a BTG($50 \times 2 \times 100$)

3.7 Limitations of Link Analysis

conjecture 2 was true for some graphs and not true for other graphs. Web pages that started out to be both hub Web pages and authority Web pages in the early stages of the procedure Filter were soon changed into either hub Web pages or authority Web pages but not both any more at the end of iterations. conjecture 3 held ground for all graphs with the exception of general Web graphs. The idea that a Web page can be both a hub page and an authority Web page is intriguing and worth further consideration. According to *Jon Kleinberg* ([65] sept 14, 2000) "nothing in the algorithm prevents this from happening but it is not a very common phenomenon". Furthermore, we only considered in this study the influence of principal eigenvectors on the selection of hubs and authorities. If we expand the role of these secondary eigenvectors, we will have not only just a set of hubs and authorities, but a multiple sets of hubs and authorities. Further research needs to focus on the structure of the different kind of graphs studied in this chapter influence the idea of a multiple sets of hubs and authorities. We showed that hubs and authorities can be lost in some BTG more than others especially when $m << N$ or $m << M$. In order to decrease the instabilities in such cases of BTG, i.e., "topic drifting", more Web pages needs to be added to the core so the ratio of the size of the core to the size of the left or right number of Web pages will increase. Also through the concept of centrality scores between 2 graphs, we showed how similar are the different cases of BTGs that we studied. Thus, some of these cases are more similar than others and thus the original idea to consider the different BTGs. More research on graph similarity can shed more light on the different structures of the BTG's and could be helpful in improving ranking. A more in depth study of the stability (*Chung* [23]) of BTG is needed.

This chapter focused on just using links as a mean to evaluate Web pages and uncover hubs and authorities. No heuristics or any other contextual information was used to further enhance the idea of hubs and authorities. In an early study, *McBryan* [78] used searching hyperlinks based on an anchor text, in which one treats the text surrounding a hyperlink as a descriptor of the page being pointed to when assessing the relevance of that Web page. *Frisse* [37] considered the problem of document retrieval in single-authored, stand-alone works of hypertext. He proposed heuristics by which hyperlinks can enhance notions of relevance (Van Rijsbergen [129]), and hence the performance of retrieval heuristics. In recent studies (*Bharat & Henzinger* [6]; *Chakrabarti, et al.* [18, 19]), three distinct user studies were performed to help evaluate the HITS system to evaluate information found on the WWW. Each one of the studies employed additional heuristics to further enhance relevance judgments. As such, these 3 studies cannot enhance the direct evaluation of the pure link-based method described here; rather they assess its performance as the core component of a WWW search tool. For example in [18], the CLEVER system was used to create an automatic resource compilation or the construction of lists of high-quality WWW pages related to a broad search topic and the goal was to see whether the output of CLEVER compared to that of a manually generated compilation such as the WWW search service of Yahoo for a set of 26 topics. A collection of 37 users was assembled; the users were required to be familiar with the use of a Web browser, but were not experts in the topics picked. The users were asked to judge each Web page as "bad", "fair", "good" or "fantastic" in terms of their utility of learning about the topic. For approximately 31% of the topics, the evaluations of Yahoo and CLEVER were equivalent to within a threshold of statistical significance; for approximately 50% of the topics CLEVER was evaluated higher; and for the remaining 19% yahoo was evaluated higher. Many of the users of these studies reported that they used the lists as starting points from which to explore, but that they visited many pages not on the original topic lists generated by the various techniques. Of course it is hard to draw definitive conclusions from these 3 studies. We believe more insight in the structure of the graphs under consideration will help improve the success of Web algorithms, i.e., HITS and CLEVER, to refine the original concepts and make place for better understanding of hubs and authorities in a variety of topological structures. This study does not pretend to have improved finding information on the WWW. This study focuses on discovering important topological structures in Web pages and to predict the behavior of Web algorithms in such environment especially that the WWW is rich not only in information but in topological structures.

4

PageRank Algorithm Applied to Web Graphs

4.1 Introduction

Commercial search engines are a key access point to the Web and have the
difficult task of trying to find the most useful of the billions of Web pages for
each user query entered. Google 's PageRank ([12]) was an attempt to resolve
this dilemma based upon the assumptions that:

1. more useful pages will have more links to them and
2. links from well linked to pages are better indicators of quality.

The continued rise of google to its current dominant position and the prolifer-
ation of other link based algorithms seem to make an unassailable argument
for the PageRank algorithm, despite the paucity of clear cut results. Modern Web IR algorithms are probably a highly complex mixture of different
approaches, perhaps optimized using probabilistic techniques to identify the
best combination. It is not possible to be definitive about commercial search
engine algorithms, however, since they are kept secret apart from the broadest
details. In fact academic research into Web IR is in a strange situation since
research budgets and data sets could be expected to be dwarfed by those of
the commercial giants, whose existence depends upon high quality results in
an incredibly competitive market place. One research that compared the two
found that the academic systems were slightly better but the authors admit-
ted that the tasks were untypical for Web users ([50]). Nevertheless, google is
one case amongst many of search algorithms gaining from approaches and developments in information science in general and bibliometrics in particular.
Traditional information retrieval techniques can give poor results on the Web,
with its vast scale and highly variable content quality. Recently, however, it
was found that Web search results can be much improved by using the information contained in the link structure between pages. The two best-known
algorithms which do this are HITS ([64]) and PageRank ([12]). The latter is
used in the highly successful google search engine. The heuristic underlying
both of these approaches is that pages with many inlinks are more likely to

G. Meghabghab and A. Kandel: *Search Engines, Link Analysis, and User's Web Behavior*,
Studies in Computational Intelligence (SCI) **99**, 69–81 (2008)
www.springerlink.com   © Springer-Verlag Berlin Heidelberg 2008

be of high quality than pages with few inlinks, given that the author of a page will presumably include in it links to pages that s/he believes are of high quality. Given a query (set of words or other query terms), HITS invokes a traditional search engine to obtain a set of pages relevant to it, expands this set with its inlinks and outlinks, and then attempts to find two types of pages, hubs (pages that point to many pages of high quality) and authorities (pages of high quality). Because this computation is carried out at query time, it is not feasible for today's search engines, which need to handle tens of millions of queries per day. In contrast, PageRank computes a single measure of quality for a page at crawl time. This measure is then combined with a traditional information retrieval score at query time. Compared with HITS, this has the advantage of much greater efficiency, but the disadvantage that the PageRank score of a page ignores whether or not the page is relevant to the query at hand. Traditional information retrieval measures like TFIDF ([4]) rate a document highly if the query terms occur frequently in it. PageRank rates a page highly if it is at the center of a large sub-web (i.e., if many pages point to it, many other pages point to those, etc.). Intuitively, however, the best pages should be those that are at the center of a large sub-web relevant to the query. If one issues a query containing the word "Clinton", then pages containing the word "Clinton" that are also pointed to by many other pages containing "Clinton" are more likely to be good choices than pages that contain "Clinton" but have no inlinks from pages containing it. The PageRank score of a page can be viewed as the rate at which a surfer would visit that page, if it surfed the Web indefinitely, blindly jumping from page to page. A problem common to both PageRank and HITS is topic drift. Because they give the same weight to all edges, the pages with the most inlinks in the network being considered tend to dominate, whether or not they are the most relevant to the query. Haveliwala ([49]) proposes applying an optimized version of PageRank to the subset of pages containing the query terms, and suggests that users do this on their own machines. We first describe PageRank. We then introduce our topology-dependent, version of PageRank, and if modified properly, it can work efficiently. Imagine a Web surfer who jumps from Web page to Web page, choosing with uniform probability which link to follow at each step. In order to reduce the effect of dead-ends or endless cycles the surfer will occasionally jump to a random page with some small probability b, or when on a page with no out-links. To reformulate this in graph terms, consider the Web as a directed graph, where nodes represent Web pages, and edges between nodes represent links between Web pages. Let W be the set of nodes, $N = |W|$, F_i be the set of pages page i links to, and B_i be the set pages which link to page i. For pages which have no outlinks we add a link to all pages in the graph. In this way, rank which is lost due to pages with no outlinks is redistributed uniformly to all pages. If averaged over a sufficient number of steps, the probability the surfer is on page j at some point in time is given by the formula:

$$PR(p_j) = \frac{(1-b)}{N} + d \sum_{p_i \in B_j} \frac{PR(P_i)}{F_i} \tag{4.1}$$

PageRank is a simple and intuitive algorithm, yet admits a mathematical implementation that scales to the billions of pages currently on the Web.

4.2 Page Rank Algorithm Applied to the Graphs of 2.2

Let r_p is the rank of a page p and od_i is the out-degree of node i, i.e., the number of outgoing links from a Web page i. The rank of a page p is computed as specified by google's search engine:

$$r_p = (1-c) + c \left[\frac{r_1}{od_1} + \frac{r_2}{od_2} + \frac{r_3}{od_3} + \ldots + \frac{r_n}{od_n} \right] \tag{4.2}$$

for all nodes 1,2,3,$\cdots$,n where $(1,p) \in R$, $(2,p) \in R$,$(3,p) \in R$,$\cdots$, $(n,p) \in R$ and where c is a constant: $0 < c < 1$ (ideally selected at 0.85). Note that the $r'_p s$ form a probability distribution over all Web pages, so the probability over all Web pages will be one.

Applying equation 4.2 to the graph in Figure 4.1 yields the following equations for all these vertices:

$$r_1 = (1-c) + c \left[\frac{r_3}{2} + \frac{r_5}{2} \right] \tag{4.3}$$

$$r_3 = (1-c) + c(r_2) \tag{4.4}$$

$$r_5 = (1-c) \tag{4.5}$$

$$r_2 = (1-c) + c \left(\frac{r_1}{3} \right) \tag{4.6}$$

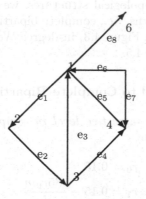

Fig. 4.1. A General Graph

Table 4.1. the values of r_1, r_2, r_3, r_4, r_5, and r_6

Rank of Vertex	Value of c	Value of i_d
$r_1 = .367$	.85	2
$r_2 = .254$	.85	1
$r_3 = .366$	.85	1
$r_4 = .47325$	.85	3
$r_5 = .15$	.85	0
$r_6 = .254$	.85	1

Replacing equations 4.3 to 4.6, in 4.2 yields:

$$r_1 = \frac{3(1-c)(2c+2+c^2)}{(6-c^3)} \tag{4.7}$$

Table 4.1 shows the values of r_1, r_2, r_3, r_4, r_5, and r_6 as a function of c: The following observations can be made:

- Vertex 4 being the vertex with the highest links pointing to it (i_d) yielded the highest ranking among all the pages.
- Vertex 1 follows vertex 4 in its ranking because of $i_d = 2$.
- Vertices 3,2, and 6 rank behind vertex 1. Although these 3 vertices have their $i_d = 1$ they should have equal value rank, yet they differ greatly. Value r_3 is closer to r_1 than it is to the rank of vertices 2 and 6, which are equal: $r_2 = r_6$.
- Vertex 5 has the least rank because it did not have any nodes pointing to it.
- Vertices with the same i_d according to equation 4.2 should yield equal ranking.

The last observation above is violated in the graph of Figure 4.1. Vertices 2, 3, and 6 should have very similar ranking. To observe whether google's ranking behaves better on other topological structures, we simulated google's algorithm on other graph patterns, *i.e.*, complete bipartite graphs like Figure 4.2. out-degree Web graphs like Figure 4.3, in-degree Web graphs like Figure 4.4, bipartite graphs like Figure 4.5.

4.2.1 PageRank Applied to Complete Bipartite Graphs

Theorem 1. *Nodes at the same tree level in complete bipartite graphs will have equal PageRank's value.*

$$r_0 = 0.15$$

$$r_1 = 0.15 + \frac{.85r_0 m}{k} \tag{4.8}$$

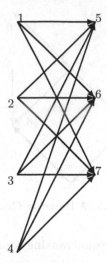

Fig. 4.2. A Complete Bipartite Graph

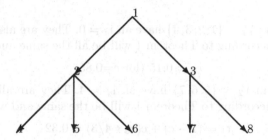

Fig. 4.3. An Out Degree Graph

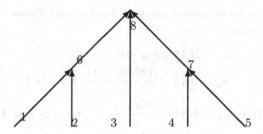

Fig. 4.4. An In Degree Graph

Where m is the number of nodes at level $= 0$, and k is the number of children to each node in Level $= 0$

Proof. Given the fact complete bipartite graphs are made of 2 levels, one level with $i_d = 0$ and another level $= 1$ with id higher than 0, then all vertices of

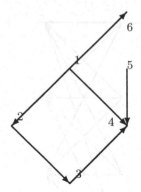

Fig. 4.5. A Bipartite Graph G_1

degree higher than 0 will have equal ranking regardless of the number of nodes or configurations.

Example: In Figure 4.2, m = 4 and k = 3. The following observations can be made:

1. All vertices in $V_1 = \{1, 2, 3, 4\}$ have an $i_d = 0$. They are also at Level = 0. Their rank according to Theorem 1 will be all the same and will equal to:

$$r_0 = 0.15 \text{ (for c=0.85)}$$

2. All vertices in $V_2 = \{5, 6, 7\}$ have an $i_d = 4$. They are all at Level = 1. Their rank according to Theorem 1 will be the same and will be equal to:

$$r_1 = (1 - c) + c(r_0 * 4/3) = 0.32$$

4.2.2 PageRank Applied to Out-degree Graphs

Theorem 2. *Nodes at the same sub-tree level in out-degree Trees will have equal PageRank's value.*

$$r_n = 0.15 \text{ for n} = 0$$
$$r_n = 0.15 + \frac{0.85(m - 1)}{m} \quad \text{for n} = 1 \qquad (4.9)$$
$$\text{One Sub-Tree}: r_n = 0.15 + \frac{.85 r_{n-1}}{m} \quad \text{for } n > 1 \text{ and } m \neq k$$
$$\text{One Sub-Tree}: r_n = 0.15 + \frac{.85 r_{n-1}}{k} \quad \text{for } n > 1 \text{ and } m = k$$

Where m is the number of children of node n-1 in one sub-tree, k is the number of children of node n-1 in the other sub-tree, and L is the level of the node

Proof. out-degree graphs (or trees) having more than 2 levels will never provide equal ranking for vertices with equal degree. Instead of the degree of a node, the level of a node in a given tree will be the determining factor.

Example. Consider the following out-degree tree in figure 4.3.

1. $r_0 = 0.15$
2. $r_1 = 0.15 + 0.85(r_0/2) = 0.214$
3. for n = 2, we have 2 ranks because $k \neq m$ (m = 3, k = 2):
    $r_2 = 0.15 + 0.85(r1/3) = 0.221$
    $r_2 = 0.15 + 0.85(r1/2) = 0.241$

Thus the rank of node 1 is r_0, the rank of nodes 2 and 3 is the same r_1, the rank of nodes 4, 5, 6 is the same $r_2 = 0.221$, and the rank of nodes 7 and 8 is the same $r_2 = 0.241$.

Interpretation: Even though all vertices are of the same degree $i_d = 1$ except the root of the tree which has a degree $i_d = 0$, vertices 2, and 3 at Level = 1 will have equal rank which is 0.214. Vertices at level = 2 will be divided into 2 ranks because the left sub-tree has more children coming out of node 2 than those coming out of node 3 on the right sub-tree. Vertices 4, 5, 6, which are at level 2 of the let sub-tree will have lower rank than those on the right part of the tree, which are still higher than that of those vertices at level = 1. This is expected since, what feeds into vertices 4, 5, 6 are nodes at a degree higher than those feeding into vertices 2 and 3, which is node 1. If 1 cites 2 and 3, and 2 now cites three others like 4, 5, and 6, then 4, 5, and 6 will have higher ranking than 2 because they are more important than 2.

4.2.3 PageRank Applied to In-degree Graphs

Theorem 3. *Nodes at the same tree level in in-degree Trees will have equal PageRank's value.*

$$r_0 = 0.15$$
$$r_n = 0.15 + .85(mr_{n-1} + kr_{n-2}) \qquad (4.10)$$

Where m is the number of parents to node n from level n-1 and k is the number of parents to node n from level n-2.

Proof. in-degree graphs (or trees) having more than 2 levels will never provide equal ranking for vertices with equal in-degree. Instead of the in-degree of a node, the level of a node in a given tree will be the determining factor.

Example. Consider the following general in-degree tree in figure 4.4.

1. $r_0 = 0.15$
2. $r_1 = 0.15 + 0.85(2r_0) = 0.405$
3. $r_2 = 0.15 + 0.85(2r_1 + r_0) = 0.555$

Thus the rank of node 8 is r_2, the rank of nodes 6 and 7 is r_1, and the rank of nodes 1, 2, 3, 4, and 5 is r_0. Interpretation: Even though vertices 6, 7, and 8 have the same degree $i_d = 2$, only vertices 6, and 7 at Level = 1 will have

Table 4.2. Google's ranking applied to Figure 4.5

Rank of Vertex	Value of c	Value of i_d
$r_1 = .15$	.85	0
$r_2 = .21$	.85	1
$r_3 = .33$	.85	1
$r_4 = .62$	.85	3
$r_5 = .15$	.85	0
$r_6 = .21$	.85	1

equal rank which is 0.405. Vertex 8, which is at level 2, will have the highest rank that is 0.555. Vertices 1, 2, 3, 4, and 5, which are at level 0, will have equal rank, which is much lower than all of other vertices. This is expected since what feeds into vertex 8 are not only a low level node such as 3 but also vertices 6 and 7, which are nodes at an in-degree higher than those feeding into 6 and 7 which are vertices 1, 2, 4, and 5 which occupy level = 0. If 1 and 2 point to 6, 4 and 5 point to 7, 6 and 7 point to 8 besides 3, that makes 8 a more important link than all other links in the tree.

4.2.4 PageRank Applied to Bipartite Graphs

Theorem 4. *PageRank's Algorithm will not work on bipartite graphs.*

Proof. Even though we have seen that bipartite graphs allow the separation of the set of vertices into 2 sets, yet nothing has changed in the interpretation of the fact that nodes with equal in-degree id can have different rankings, which is contrary to the interpretation of equation 4.1.

Example. Consider the example of a bipartite graph like the one in figure 4.5.

Table 4.2 summarizes google's ranking applied to Figure 4.5. The same remark that was applied to the graph in figure 4.1 still applies to the graph in Figure 4.3. Vertex 3 has a higher ranking than those of the 3 vertices 2, 3, and 6, which have the same in-degree. Making a graph a bipartite graph did not change anything in the ranking of links. A more involved ranking scheme will be needed to further explore the complexity of ranking in bipartite and General graphs.

4.2.5 PageRank Applied to General Graphs

Theorem 5. *PageRank's Algorithm will not work on General graphs.*

Table 4.3 summarizes the results of the modified google's ranking procedure on these different topological structures. Thus such an algorithm needs to be applied carefully depending upon the topological structure of the Web pages that are studied. No other studies have reported the influence of the topology of the Web links on google's Page Ranking technique.

Table 4.3. Modified google's ranking

Topological Structure	google's Web ranking
in-degree Trees	Works well per Tree level(Theorem 3)
out-degree Trees	Works well per Sub-Tree level(Theorem 2)
Bipartite Graphs	Does not work (Theorem 4)
General graphs	Does not work (Theorem 5)
complete bipartite graphs	Works well per graph level (Theorem 1)

4.2.6 Summary of the Application of the PageRank Algorithm to the Graphs of 2.2

Did the classification of the web graphs of 2.2 developed in equation 2.11 in section 2.5 of chapter 2 help shed a light on the behavior of PageRank's algorithm ([12]) even though PageRank's algorithm is not related to the (Out-Degree, In-degree) space in which all these different graphs have been studied. According to equation 2.11 complete bipartite graphs(CBG) have a behavior close to out-degree graphs(OD), and that OD graphs have a behavior close to general graphs(GG), and that GG have a behavior close to in-degree graphs (ID). PagerRank algorithms works well per graph level for CBG, works well for OD per Sub-Tree level, Works well per Tree level for ID, but does not work not for GG and BG. According to 2.11 PageRank algorithms should not have held ground for ID given the fact that their behavior is close to GG than the other graphs. New classification schemes are needed to help predict the behavior of PageRank algorithms in the new space. The application of PageRank algorithm on Bow tie graphs will be considered next

4.3 PageRank Algorithm Applied to Bow Tie Graphs

As seen in Figure 4.6, a BTG is made out of 2 graphs, a left graph which is a bipartite graph or BG(L) which includes all the nodes from 1,2,3,···,10 which are the origination nodes, and node 11 which is the core; and a BG(R) which is made out of node 11, and nodes 12,13,···,21. In section 2.3, we established the relationship between a BTG and the left complete bipartite graph and the right complete bipartite graph. We established that the adjacency matrix of the BTG is made out of the adjacency of the left LBG and the right CBG. The number of inlinks of the BTG is made out of the inlinks of the CBG(L) and the CBG(R). The number of outlinks of the BTG is made out of the outlinks of the CBG(L) and the number of outlinks of the CBG(R). The latter was a special case to begin the study of BTG's. In this section, we start out with a BTG (N*m*N) as a set of 2 complete bipartite graphs CBG with $m << N$, a left one CBG(L) and a right one CBG(R). They share a core in the middle as seen in figure 4.6. Example: Applying PageRank to

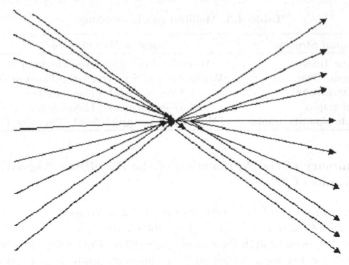

Fig. 4.6. A 10*1*10 bow tie graph

Table 4.4. Ranking of the nodes of Figure 4.6

Rank of Vertex	Value of c	Value of i_d
$r_{1,2,3,4,5,6,7,8,9,10} = .15$	.85	0
$r_{11} = 1.425$	.85	10
$r_{12,13,14,15,16,17,18,19,20,21} = .2775$	.85	1

a BTG(10*1*10): The BTG of Figure 4.6 is made out of 2 CBG(L) and a CBG(R). We use equation 4.8 and applied for 3 levels:

1. At level L = 0, the ranking of nodes 1, through 10 is : $r_0 = 0.15$
2. At level L = 1, using 4.8: $r_1 = .15 + .85(mr_0/k) = 1.425$
3. At level L = 2, using 4.8: $r_2 = .15 + .85(mr_1/k) = .2775$.

As seen in table 4.4, nodes at level 0, 1 through 10, have the lowest rank. Nodes at the core which is 11 has the highest rank, while the nodes at level 2, 12 though 20, have a rank higher than nodes of level 0 but lower than level 1. It seems like google's ranking is working. Nodes with the highest i_d have the highest rank, while the nodes with lowest i_d have the lowest rank. We consider the following cases:

1. A general bow tie graph: BTG(N*m*M) with $m << N$ and $m << M$, with a CBG(L) and a CBG(R) but where $m << N$ and $m << M$ and we consider the cases of $N > M$ and $N < M$. In the case of $N < M$, we find that:

   for Level L = 0;
   $r_0 = 0.15$ for all vertices from 1 through N
   For level L = 1, using 4.8:

Table 4.5. Google's ranking on a general bow tie graph

Rank of Vertex	Value of c	Value of i_d
$r_{1..N} = .15$	.85	0
$R_{N+1,.N+m} = N/10 * m > .15$	.85	N
$R_{N+m+1..N+M} = .15$	.85	$0 << m << N$

Table 4.6. Google's ranking does work in the case of $N > M$

Rank of Vertex	Value of c	Value of i_d
$r_{1..N} = .15$	.85	0
$R_{N+1,.N+m} = N/10 * m > .15$	.85	N
$.15 < R_{N+m+1..N+M} < N/10 * m$	.85	$0 << m << N$

$r_1 = 0.15 + 0.85(r_0 * m/k)$.
$r_1 = .15 + .85*(.15*m/k) = .15 + .85*(.15*N/m)$
$r_1 = .15 + N/m^* 10 \approx N/10*m$ for all vertices from N+1 through N+m.
For Level L = 2, using 4.8:
$r_2 = .15+0.85(r_1 * m/k)=.15+.85*N*m/10*m*M$
$r_2 = .15+.85*N/10*M$
$r_2 = \approx .15$ for all vertices from N+m+1 through N+M

According to table 4.5, then although the i_d of all nodes from N+m+1 through N+M is $m > 0$, yet the rank of all these nodes is really very close to .15. Thus google's ranking does not work in that case.
In the case of $N > M$, we find that (see table 4.6):

for Level L = 0;
$r_0 = .15$ for all vertices from 1 though N
For Level =1, using equation 4.8:
$r_1 = .15 + .85*(.15*m/k) = 15 + .85*(.15*N/m) = .15 + N/m^* 10$
$r_1 \approx N/10*m$
For Level =2, using 4.8:
$r_2 = .15 + 0.85(r_1 * m/k)=.15+.85*N*m/10*m*M$
$r_2 = .15 + .85*N/10*M > .15$

It appears that google's ranking works in this case.

2. A general bow tie graph: BTG(N*m*M) with $m << N$ and $m << M$, as made out of 2 graphs: a complete BG on the left CBG(L) and bipartite graph on the right BG(R) or a BG(L) and CBG(R), which are reasonable cases to study and with $N > M$ or $N < M$. If $N < M$ or $N > M$, and a CBG(L) and a BG(R), we find that:

We still use equation 4.8 for the CBG(L): $R_0 = .15$ for all nodes 1..N
$R_1 = .15 + .85*.15*N/m = .15 + N/10m$
For BG(R), we have a bipartite graph and we know that in that case

according to Theorem 4 that google's ranking does not work. If we use a BG(L) and a CBG(R), with $N > M$ or $N < M$, we know that google's ranking will not work on the left but will work on the right.

3. A general bow tie graph: BTG(N*m*M) with $m < N$ and $m < M$, as made out of 2 graphs: a CBG(L) and BG(R) or a BG(L) and CBG(R), which are reasonable cases to study and with $N > M$ or $N < M$. Case 3 follows Case 2 since we have a BG(R) one time and a BG(L) another time. In case 3. in general we conclude that google's ranking will not work.

4. A general bow tie graph where $m >> M$ and $m >> N$ as made out of 2 graphs: a CBG(L) and a CBG(R). In the case of $N < M$ or $N > M$, we find that:

for level L = 0

$r_0 = .15$ for all vertices from 1 through N

For level = 1, using 4.8 :

$r_1 = .15 + 0.85(r_0 * m/k)$

$r_1 = .15 + .85*(.15*m/k) = .15 + .85*(.15*N/m) = .15 + N/m * 10$

$r_1 \approx .15$ for all vertices from N + 1 through N + m

For level = 2, using 4.8:

$r_2 = 0.15 + 0.85(r_1 * m/k) = .15 + .85 * .15 * m/M$

$r_2 = .15 + .1269 * m/M$

$r_2 > .15$ for all vertices from N + m + 1 through N + M

According to table 4.7, then although the i_d of all nodes from N+1 through N+m is $N > 0$, yet the rank of all these nodes is really very close to .15 which is equal to the ranking of the nodes with $i_d = 0$. Thus google's ranking does not work in that case.

5. A general bow tie graph where m = M and m = N as made out of 2 graphs: a CBG(L) and a CBG(R) In that case, we find that:

for Level=0

$r_0 = 0.15$ for all vertices from 1 through N

For level = 1, using 4.8:

$r_1 = .15 + 0.85(r_0 * m/k)$

$r_1 = .15 + .85*(.15* m/k) = .15 + .85*(.15*N/N)$

$r_1 = .15 + N/N * 10 = .25$ for all vertices from N+1 through N+m

Table 4.7. Google's ranking for a bow tie graph where $N < M$

Rank of Vertex	Value of c	Value of i_d
$r_{1..N} = .15$	.85	0
$R_{N+1,.N+m} = N/10 * m > .15$	.85	N
$R_{N+m+1..N+M} > .15$	.85	$0 < m << N$

Table 4.8. Google's ranking of a bow tie graph where m = M and m = N

Rank of Vertex	Value of c	Value of i_d
$r_{1..N} = .15$	.85	0
$R_{N+1,.N+m} = .25$	.85	N
$R_{N+m+1..N+M} > .3625$	.85	N

For level = 2, using 4.8:

$r_2 = .15 + 0.85(r_1 * m/k) = .15 + .85 * .25 * m/M$

$r_2 = .15 + .2125 = .3625$ for all vertices from N+m+1 through N+M.

According to table 4.8, then although the i_d of all nodes from 2N+1 through 3N is N, yet the rank of all these nodes is .3625 which is greater than .25 for all the nodes of N+1 thought 2N which have the same i_d. Google's ranking does not work in that case.

4.4 Limitations of PageRank Algorithm

Google's ranking algorithm need to be adjusted to different topological Web structures to be able to successfully rank Web pages without any contextual information added. Given the fact google's search engine is the forefront indexing search engine with more than "eight" billion Web pages and being adopted by major search engines like Yahoo it seems that a study of their ranking Web algorithm is timely to further explore its applicability in a variety of Web graphs. This study focused first on categorizing different Web graphs and classifying them according to their in-degree out-degree properties. Then we applied google's Web ranking algorithm to the complete bipartite graphs, followed by bipartite graphs, then out-degree trees, in-degree trees, and last General graphs. Google's ranking Web algorithm worked best on complete bipartite graphs by ranking equally vertices at the same level. The algorithm did not fare well in other Web graph structures on the lower ranking of the remaining vertices. More specifically, vertices with equal degrees, e.g., equal amount of outgoing nodes did not rank equally. Different theorems were adopted for these different topological structures, and readjusted google's ranking to fit these different topological structures. Other ranking algorithms could have been studied and applied to the different topological Web graphs, but given the popularity of google, its wide indexing power, more than a billion Web pages, makes it a very powerful search engine that has been adopted by other search engines like Yahoo. All of these factors contributed to the consideration of google's ranking algorithm in this study.

Stochastic Simulations Of Search Engines

5.1 Introduction

Computer simulations are widely used in a variety of applications including the
military, service industries, manufacturing companies, nuclear power plants,
transportation organizations, and manufacturing domain. For example, in nuclear power plants, often computer simulations are used to train personnel on
failures and normal operation, study operation plans, support testing of the
heart of the nuclear power plant, as well as to evaluate and compare future
design plan changes. Other techniques that are used to examine systems, in
general, do not have the advantages that computer simulations bring mainly
because computer simulations provide cheaper and more realistic results than
other approaches. In some cases, computer simulation is the only means to
examine a system, like in nuclear power plants, since it is either too danger-
ous to bring a system under such failure conditions or too costly especially for
combat situations to experiment with the system. Also, computer simulations
permit us to study systems over a long period of time, to learn from real
world past experiences, and to have control over experimental conditions. Actual Internet simulation models are expensive to develop and use in terms of
personnel, time and resources. Large memory requirements and slow Response
time can prevent companies from considering it as a useful tool. Developing
Internet simulations models during training requires models that are very
close to reality while their speed is secondary. During testing, speed and re-
producibility become the primary issues, which incite us to make the different
internal simulation modules as efficient and accurate as possible. Computer
simulations have provided companies with the description of the input settings that are needed to produce the optimal best output value for a given
set of inputs in a specific domain of study. Response surface methodologies
using regression models approximations of the computer simulation were the
means to achieve computer simulation optimization. As *Myers et al.* ([104])
stated it in Technometrics,

G. Meghabghab and A. Kandel: *Search Engines, Link Analysis, and User's Web Behavior*,
Studies in Computational Intelligence (SCI) **99**, 83–104 (2008)
www.springerlink.com   © Springer-Verlag Berlin Heidelberg 2008

"There is a need to develop non-parametric techniques in response surface methods (RSM). The use of model-free techniques would avoid the assumption of model accuracy or low-level polynomial approximations and in particular, the imposed symmetry, associated with a second degree polynomial (page 139)."

One possible non-parametric approach is to use artificial neural networks (ANN).

5.1.1 Artificial Neural Networks (ANN) on the WWW

An ANN learns to imitate the behavior of a complex function by being presented with examples of the inputs and outputs of the function. This operation is called training. A successfully trained ANN can predict the correct output of a function when presented with the input (or inputs for a multivariate function). ANN has been applied successfully in a variety of fields including industrial and mining and manufacturing, business and finance and marketing, medicine and health, an up to date Web page of the application of neural networks to medicine (http://www.phil.gu.se/ANN/ANNworld.html). Searching for information, in general, is a complex, fuzzy, and an uncertain process. Information retrieval models have been applied to online library catalogs in the past and these models can be still used for any online information service. The WWW is a system of storage and retrieval of information characterized by its enormous size, hypermedia structure, and distributed architecture. Computer and information scientists are trying to model the behavior of users and examine their successes and failures in using search engines on the WWW. While understanding user information seeking behavior on the Web is important, evaluating the quality of search engines is an important issue to tackle. WWW users will be more successful if they were made aware of how to interpret their search results when they query the WWW, and how these results are linked directly to the search engine they use. ANN has been used to help users negotiate their queries in an information retrieval system to better assess their subject needs. *Meghabghab & Meghabghab* ([82]) designed an ANN to act as an electronic information specialist capable of learning to negotiate a patron's query and translate it into a true, well formulated statement prior to accessing an online public access catalog (OPAC). *Meghabghab & Meghabghab* ([99]) proposed a new neural network paradigm called "Selection Paradigm" (SP) to solve the problem of algorithm selection in multi-version information retrieval. SP learns to choose a preferred pattern for a given situation from a set of n alternatives based on examples of a human expert's preferences. The testing results after training are compared to Dempster-Shafer's (DS) model. DS computes the agreement between a document belief and two or more reformulated query beliefs by two or more different algorithms. Final testing results indicate that SP can produce consistent rank orderings and perform better in certain information retrieval situations than DS's model. More recently, connectionist methods such as fuzzy clustering and Kohonen Self-Organizing Maps (SOM) were also used to evaluate log data analysis in a geo-referenced information service of a digital library. There are not many examples of ANN being utilized in the evaluation of the quality of search engines. This study is a premiere not only in using neural networks in the evaluation of search engines, but also in comparing its results to other simulation techniques that have been used in other applications. In the next section, we will apply ANN in approximating computer simulations of the quality of search engines.

5.1.2 Rejected World Wide Web Pages

Real-world queries that is, queries that have been used in real world situations, are the only way for testing the WWW as to estimate the relative Coverage of search engines with respect to existing Web pages. This can be a lot different from the relative number of pages indexed by each search engine. Indexing and retrieval vary from one engine to another. For example, there may be different limits on the size of pages indexed by an engine, the words that are indexed, and processing time per query. google has a lexicon of 14 million words and a limit of 1 to 10 seconds per query (dominated by disk Input/Output over UNIX File System). Once 40,000 matching documents are found for a given query (to put a limit on Response time), sub-optimal results are returned not based on the user needs or query formulation but how well the pages are indexed. If all pages were to be equally indexed, information access across all Web pages would be equal. But designers tend to choose Web pages to be indexed. The Web page of President Bill Clinton, for example, will be indexed before the Web page of Joe Doe in Arkansas. The bias of indexing is complicated with the decision to crawl. The percentage of crawling versus indexing is a challenge because of costs of hardware and limitations of software (operating systems). Crawling is important in the beginning for a new search engine to cover as many Web pages as possible to attract users to come back, but then it tapers off as the cost to keep updating databases keeps increasing. Web pages are "living communities" of hyperlinks (*Kumar et al.* [70]). They are updated, modified, deleted, and changed constantly. The nature of Web pages complicates the evaluation of the returned set of results of search requests. To date, studies of Web search engines have mainly focused on retrieval performance and evaluation of search features. No studies have evaluated the quality of search engines based on the proportion of Web pages rejected and, therefore, excluded from search results. We investigate the proportion of rejected Web pages for a set of search requests that were submitted to 4 search engines: Alta Vista, google, Northern Light, and yahoo. We applied two techniques to the resulting set of data: RBF neural networks, and second order multiple linear regression models.

5.2 ANN Meta-model Approach for the WWW

A meta-model is a model of a model. Typical simulation meta-modeling approaches involve the use of regression models in response surface methods. A few attempts have been made to employ neural networks as the meta-modeling technique. *Pierreval & Huntsinger* ([110]) were successful in using ANN as meta-models of stochastic computer simulations. The common feature of these models was an ANN baseline, which involves using a back-propagation, trained, multi-layer ANN to learn the relationship between the simulation inputs and outputs. The baseline ANN meta-model approach was developed on an (s, S) inventory computer simulation and was applied to a larger application in the domain under consideration. RBF neural networks are emerging as a viable architecture to implement neural network solutions to many problems. RBF neural networks are deterministic global non-linear minimization methods. These methods detect sub-regions not containing the global minimum and exclude them from further consideration. In general, this approach is useful for problems requiring a solution with guaranteed accuracy. These are computationally very expensive. The mathematical basis for RBF network is provided by Cover's Theorem (*Cover* [25]), which states:

"A complex pattern-classification problem cast in a high dimensional space nonlinearly is more likely to be linearly separable in a low-dimensional space".

RBF uses a curve fitting scheme for learning, that is, learning is equivalent to finding a surface in a multi-dimensional space that represents a best fit for the training data. The approach considered here is a generalized RBF neural network having a structure similar to that of Figure 5.1. A generalized RBF has the following characteristics:

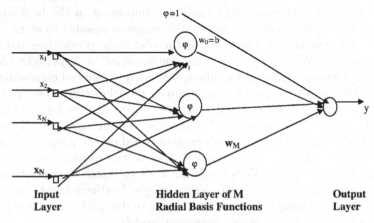

Fig. 5.1. A generalized RBF

1. Every radial basis function ϕ_i at the hidden layer has a center position t_i and a width or a spread around the center. It connects to the output layer through a weight value w_i.
2. The number of radial basis functions at the hidden layer is of size M, where M is smaller than the number of input training patterns N.
3. The linear weights w_i which link an RBF node to the output layer, the position of the centers of the radial basis functions, and the width of the radial basis functions are all unknown parameters that have to be learned.

A supervised learning process using a Gradient Descent (GD) procedure is implemented to adapt the position of the centers and their spreads (or widths) and the weights of the output layer. To initialize GD, the authors begin the search from a structured initial condition that limits the region of parameter space to an already known useful area by using a standard pattern-classification method . The likelihood of converging to an undesirable local minimum in weight space is already reduced. Also, a supervised learning process using Interior Point Method (IPM) is implemented to adapt the position of the centers and their spreads (or widths) and the weights of the output layer, but which reduces the amount of computation compared to the one developed with GD. A standard Gaussian classifier is used; it assumes that each pattern in each class is drawn from a full Gaussian distribution.

5.2.1 Training RBF without Correction of Centers, Spreads and Weights (RBF without CCSW)

The initial step in training using RBF is to pursue a maximum likelihood method, which views parameters as fixed but unknown. To be more specific, suppose that we have a two-class problem (it could be generalized to a multi-class case), with the underlying distribution of the input observation vector is Gaussian. The vector of observation or training pattern x is then characterized as follows:

$x \in X_1$: Mean vector = μ_1 Covariance matrix = C
$x \in X_2$: Mean vector = μ_2 Covariance matrix = C

where $\mu = \mathrm{E}[\mathrm{X}]$, $C = E[(x - \mu)(x - \mu)^T]$, and E denotes the statistical expectation operator or mean.

The input-output relation of the network is defined by:

$$y(x) = \sum_{i=1}^{M}(w_i\phi_i(x) + b) \qquad (5.1)$$

where $\phi_i(x)$ is a radial basis function, and b is the bias. A radial basis function has a center t_i and is defined as follows:

$$\phi_i(x) = G(||x - t_i||_C), i = 1, 2..M \qquad (5.2)$$

where G is a Gaussian function centered at t_i with a general weighted norm defined as follows:

$$||x||_C = (Cx)^T(Cx) = x^T C^T C x \tag{5.3}$$

where C is norm weighting square matrix having the same dimension as the input vector x. Replacing 5.2 in 5.1 yields:

$$y(x) = \sum_{i=1}^{M}(w_i G(|x - t_i|_C) + b \tag{5.4}$$

Where M is the number of centers that belong to class X_1. To fit the training data with the desired output, we require:

$$y(x_j) = d_j \text{where j=1,.....N} \tag{5.5}$$

Where x_j is an input vector and d_j is the desired output. Let:

$$g_{ji} = G(||x_j - t_i||_C) \tag{5.6}$$

Where i = 1,...M and j = 1,.....N. $G(||x_j - t_i||_C)$ can be expressed as follows:

$$G(||x_j - ti||_C) = \exp[-(x - t_i)^T C^T C(x - t_i)] \tag{5.7}$$

Define $\sum^{-1}$ as $(1/2)\sum^{-1} = (C_i^T C_i)$, equation 5.7 yields:

$$G(||x_j - t_i||_C) = \exp[-(1/2)(x - t_i)^T \sum^{-1}(x - t_i) \tag{5.8}$$

Equation 5.8 represents a multivariate Gaussian distribution with mean vector t_i and covariance matrix $\sum$. In a matrix form, G can be written as $G = [g_{ji}]$, where $g_{ji} = [g_{j1}, g_{j2}, ..., g_{jM}, 1]^T$ and where j = 1,........,N and i = 1,.......M. Combining equations 5.6 and 5.5, and defining w = $[w_1, w_2, w_3, w_4, ..., w_M]^T$, equation 5.4 can be written in a matrix format as

$$Gw = d \tag{5.9}$$

To inverse 5.9 we have to use a pseudo-inverse where:

$$w = G^+ d = (G^T G)^{-1} G^T d \tag{5.10}$$

where $G^T G$ is a square matrix with a unique inverse of its own.

5.2.2 Training RBF with Correction of Centers, Spreads, and Weights using Gradient Descent (RBF with CCSW using GD)

For the training of the proposed network, we used the modified generalized delta rule to update the output weights as well RBF centers and spreads.

We train the network by minimizing the sum of the squares of the errors for all output neurons. Let E(n) be the sum of the squares of the errors for all output neurons at time n:

$$E(n) = (1/2) \sum_{j=1}^{N} (e_j(n))^2 \tag{5.11}$$

Where $e_j(n)$ is the error signal of output unit defined at time n and is defined by:

$$e_j(n) = d_j(n) - y(x_j) = d_j(n) - \sum_{j=1}^{M} w_i(n) G(\|x_j - t_i(n)\|_C) \tag{5.12}$$

The problem is to find the free parameters w_i, t_i, and $(\sum^{-1})$ to minimize E. The results of the minimization can be summarized as follows for the linear weights of the output layer:

$$\frac{\partial E(n)}{\partial w_i(n)} = \sum_{j=1}^{N} e_j(n) G(\|xj - ti\|_C) \tag{5.13}$$

And:

$$w_{i+1}(n) = w_i(n) - \zeta_1 \frac{\partial E(n)}{\partial w_i(n)} \quad i = 1, 2.....M \tag{5.14}$$

and for the positions of centers at the hidden layer:

$$\frac{\partial E(n_i}{\partial t_i(n)} = 2w_i(n) \sum_{j=1}^{N} e_j(n) G'(\|x_j - t_i(n)\|_C) \sum_{i}^{-1} (x_j - t_i(n)) \tag{5.15}$$

where G' is the derivative of G with respect to its argument.

$$t_{i+1}(n) = t_i(n) - \zeta_2 \frac{\partial E(n)}{\partial t_i(n)} \quad i = 1, 2.....M \tag{5.16}$$

For the spreads of centers at the hidden layer:

$$\frac{\partial E(n_i}{\partial \sum_i^{-1}(n)} = w_i(n) \sum_{j=1}^{N} e_j(n) G'(\|x_j - t_i(n)\|_C) Q_{ji}(n) \tag{5.17}$$

where $Q_{ji}(n) = [x_j - t_i(n)][x_j - t_i(n)]^T$

$$\sum_i^{-1}(n+1) = \sum_i^{-1}(n) - \zeta_3 \frac{\partial E(n)}{\partial \sum_i^{-1}(n)} \quad i = 1, 2.....M \tag{5.18}$$

Where ζ_1, ζ_2, and ζ_3 are different learning rate parameters that could be made adaptative themselves.

5.2.3 Training RBF with Correction of Centers, Spreads and Weights using Interior Point Methods (RBF with CCSW using IPM)

To evaluate the promise of the IPM for RBF, we trained the centers, spreads and the weights as follows. The learning error function E can be approximated by a polynomial P(x) of degree 1, in a neighborhood of xi :

$$E \approx P(x) \equiv E(x_i) + g^T(x - x_i) \qquad (5.19)$$

$$g = \nabla_x E = \sum_{j=1}^{N} \nabla_x(xE_j)$$

So the linear programming problem for a given search direction has as a solution:

$$\text{minimize} \quad g^T(x - x_i) \qquad (5.20)$$
$$\text{subject to} \quad -\alpha o \leq x - xi \leq \alpha o$$

Where o is the vector of all ones with the same dimension as the state vector x and $\alpha > 0$ is a constant. In this study we rely on the method applied in *Meghabghab & Ji* ([100]) and apply it three times as follows:

1. Search for the direction that will minimize the error corresponding to the weights.
2. Search for the direction that will minimize the error corresponding to the centers of the neurons.
3. Search for the direction that will minimize the error corresponding to the spreads of the neurons.

The linear interior point method and its dual can be expressed as follows:

$$\begin{array}{llll}
\text{minimize} & g^T x & \text{maximize} & b^T y \\
\text{subject to} & x + z = b & \text{and} \quad \text{subject to} & s_x = c - y \geq 0 \\
& x, \, z \geq 0 & & s_z = -y \geq 0.
\end{array} \qquad (5.21)$$

Where $x - x_i + \alpha o = u$ and $b = 2\alpha o$.

The vector x could represent the weights of the neurons, the centers of the neurons, or the spreads of the neurons. Thus (x, z, y, s_x, s_z) is an optimal feasible solution if and only if:

$$z = b - x$$
$$s_x = c - y$$
$$s_z = -y$$
$$Xs_x = 0 \qquad (5.22)$$
$$Zs_z = 0$$

Where X is a diagonal matrix with diagonal elements made out of the values of vector x and Z is a diagonal matrix with diagonal elements made out of the values of vector z. Given $(x^k, z^k, y^k, s_x^k, s_z^k)$ which is strictly feasible we apply Newton's method to the perturbed non linear method of 5.22 resulting in the following system of equations:

$$\delta x = (X^{-1}S_x + Z^{-1}S_z)^{-1}[Z^{-1}(Zs_z - \mu e) - X^{-1}(Xs_x - \mu e)]$$
$$\delta z = -\delta x$$
$$\delta s_x = -X^{-1}(S_x\delta x + Xsx - \mu e)$$
$$\delta s_z = Z^{-1}(S_z\delta x - Zs_z - \mu e) \tag{5.23}$$
$$\delta y = -\delta s_x$$

Now we set:

$$(x^{k+1}, z^{k+1}, y^{k+1}, s_x^{k+1}, s_z^{k+1}) = (x^k, z^k, y^k, s_x^k, s_z^k) + \theta^k(\delta x, \delta z, \delta y, \delta s_x, \delta s_z), \tag{5.24}$$

The choices of μ and θ_k determine different interior point algorithms. We define with the judicial choice of μ and θ_k an algorithm with polynomial complexity.

5.2.4 New RBF Interior Point Method Learning Rule

In our problem an initial point is easily obtained by exploring its special structure. We choose:

$$x^0 = \alpha e, z^0 = \alpha e, sz^0 = \rho e, y^0 = -\rho e, sx^0 = c - \rho e \tag{5.25}$$

where $\rho = max_{1 < i < n}\{\alpha, max\{\alpha - c_i\}\}$. Clearly $(x^0, z^0, y^0, sx^0, sz^0)$ is a strictly feasible point to 5.21.

Procedure 1

P_1 Initially set $\rho_{min} = (2n)^{1.5}$, $\beta = 0.9995$, and choose $(x^0, z^0, y^0, sx^0, sz^0)$ from equation 5.25

P_2 For k = 0,1,2

P2.1 $\rho_k \leftarrow max\{\rho_{min}, 1/((x_k)^T sx_k + (z_k)^T sz_k)\}$

P2.2 Calculate the step $(\delta x, \delta z, \delta y, \delta s_x, \delta s_z)$ from 5.22

P2.3 Set $\theta_1^k = max\{\theta | \theta \leq 1, x^k + \theta\delta x \geq 0, z^k + \theta\delta z \geq 0, s_x^k + \theta\delta s_x \geq 0,$ $s_z^k + \theta\delta s_z \geq 0$

P2.4 If $\theta_1^k \leq 0.5$, then divide ρ_{min} by 2 and set $\theta^k = \beta\theta_1^k$

P2.5 If $0.5 \leq \theta_1^k < 1$, then set $\theta^k = \beta\theta_1^k$

P2.6 If $\theta_1^k = 1.0$ then multiply ρ_{min} by 2 and set $\theta^k = 1$

P2.7 Update the iterate from 5.24 and go to the next iteration.

A new learning rule based on Procedure 1 can now be specified as follows:
An interior-point learning rule applied to weights

A1. Choose w_0 and $\alpha > 0$.

A2. For k = 0,1,2,...

A2.1 Compute $\triangledown E_{avg}(w_k)$, gradient of $E_{avg}(w)$ at w_k

A2.2 Call Procedure 1, let $(x^*, z^*, y^*, sx^*, sz^*)$ be the solution to 5.21

A2.3 Update the weights by $w_{k+1} = w_k + x^* - \alpha e$,

The same interior point learning rule applied to weights can be applied to the positions of the centers of neurons and to the spreads of the neurons. *Meghabghab & Nasr* ([101]) applied the 3 RBF training algorithms to the following benchmark problems:

Sonar Problem: This problem discriminates sonar signals bounced off a metallic cylinder and those bounced between two sets of training points that lie on two distinct spirals in the x-y plane. Each spiral has 94 points.

Two-Spiral Problem: This problem discriminates between two sets of training points that lie on two distinct spirals in the x-y plane. Each spiral has 94 input-output pairs in both the training and test pairs.

Vowel Recognition Problem: This problem trains a network for speaker independent recognition of the 11 steady-state vowels of English. Vowels are classified correctly when the distance of the correct output to the actual output is the smallest among the distances from the actual output to all possible target outputs.

10-parity problem: This problem trains a network that computes the modulo-two sum of the 10-binary digits. There are 1024 training patterns and no testing set.

Figure 5.2 summarizes the results of the 3 algorithms on the four benchmark problems.

Meghabghab & Nasr ([102]) developed an iterative RBF meta-model approach to approximating discrete event computer simulations in order to provide accurate meta-models of computer simulations. An RBF neural network was used to learn the relationship between the simulation inputs and outputs.

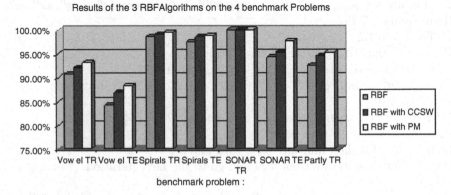

Fig. 5.2. Results of the 3 Algoritms

The iterative RBF meta-model approach starts with small training and testing sets, and builds RBF meta-models to use in performing factor screening to eliminate those input factors that do not appear to make much difference on the simulation output. After eliminating the irrelevant factors, the baseline approach could be used on the remaining factors with substantial savings in total computer simulation runs. The iterative RBF neural network was developed and used to evaluate the quality of 4 different search engines on a variety of queries.

5.2.5 RBF Training Algorithm for Web Search Engines

This study employed RBF to estimate the proportion of rejected Web pages or URL's after they have been tested for lower than the maximum allowable threshold on four search engines. In selecting Web search engines for this study, particular attention was paid for evaluating their retrieval performance. Only 4 search engines were evaluated: Alta Vista, google, Northern Light, and yahoo. Nine variables were chosen as inputs for the simulation based on studies on other search engines. These 9 variables and their denotations are:

1. Precision (X_1).
2. Overlap (X_2): common portion of the Web that is indexed among all the engines that we have selected for the study.
3. Response time (X_3).
4. Coverage (X_4): Coverage is in terms of Web pages considered.
5. Coverage (X_5)
6. Boolean logic (X_6).
7. Truncation (X_7).
8. Truncation (X_8).
9. Portion of the Web pages indexed (X_9) (e.g., titles plus the first several lines, or the entire Web page)

The output variable is the proportion of rejected Web pages or URL on a given query or search question. Such a query or search question is called an exemplar. An exemplar is a format of a query that can represent queries of the same kind regardless of the content involved in the query. An example of such an exemplar would be: "violence among teenagers". That same query could be "violence among athletes" or "drugs among teenagers" or "drugs among athletes" and would be considered of the same degree of fuzziness or vagueness. An exemplar is said to be unsuccessful or redeemed with no answers from the search engine if its proportion of rejected Web pages exceeds a certain threshold. The set of queries used to search the Web is important in studying the performance of the search engines on the WWW. Query selection to test the WWW must be different from any other set of well-controlled collection because pages on the WWW are not homogeneous Web pages. The reason that the authors did not use the TREC (Text REtrieval Conference) (see the URL: http://trec.nist.gov) queries for example, to test the Web is the fact that the Web is a collection of heterogeneous documents. The standard deviation between 2 randomly selected Web pages is so high that it is extremely difficult to apply the results of the TREC experiment to the Web. A wide number of users access the WWW, with different cultural backgrounds, different language skills, different languages, which makes it hard to express the same thing with the same vocabulary in a Web page. This does not mean that the meta-information that is extracted cANNot be homogeneous but this is generated information on the actual existing information. ZD Labs recently conducted a study among 5 search engines including google, yahoo, Direct Hit, Alta Vista and Northern Light to determine which search engine produced the most relevant results (the queries used in the study were not disclosed). AltaVista fared better than google, Direct Hit, Northern Light and yahoo. It had the most relevant results for multiword queries. google had the best results for single word queries (See the URL: http://cbs.marketwatch.com/archive/20000503/news/current/altavista).

Given the intense competition among commercial search services, finding the list of all existing URLs is highly prized proprietary information. The threshold of Web pages rejection that is adopted by different Web engine designers is even harder to determine. What is even worst is the fact that such a threshold varies among the different Web engines. The authors of this study will assume that this number does not change and is set around 80%. This number might seem high, but based on all the parameters considered during the testing by different search engines, it represents an accurate estimate of what constitutes the definition of when a exemplar is accepted or when an exemplar is rejected by a given search engine. The data used was limited to 70 different exemplars because all inputs and output were complete for these different exemplars. The numbering of these different exemplars is no secret but an easy way to reference the queries in question.

5.2.6 Regression Models Approach for the WWW ([87])

In this section, we followed the experimental design and optimization chapter on an inventory system developed by *Law & Kelton* ([72]). The mechanistic model of the simulation used Simulink (a toolbox provided by Matlab). The data used to develop the empirical approximation models consists of 2 different sets of the 9 different input variables, which will be denoted by X_1, X_2, X_3, X_4, X_5, X_6, X_7, X_8, and X_9. The first data set will consist of 29 combinations of the variables $X_1, X_2, X_3, X_4, X_5, X_6, X_7, X_8$, and X_9 with 2 different set values for each variable. The variable Precision X_1 can vary from .1 at the lowest to 1.0 at the highest, e.g., $X_1 = \{0.1, 1.0\}$. As reported by Bharat & Broder ([6]), the variable Overlap or X_2 varies from 25% which is the least with up to 35% being the maximum among all search engines, e.g., $X_2 = \{0.25, 0.35\}$. The variable Response time X_3 varies from 1 to 5 seconds, e.g., $X_3 = \{1, 5\}$. The variable Coverage at the time of simulation of the queries, varied from 250 millions Web pages which the least up to 512 millions even though the latest estimate shows google now reaching up to 1 billion Web pages in its Coverage, e.g., $X_4 = \{250000000, 512000000\}$. The variable Coverage can vary from weekly with up to 1 month, e.g., $X_5 = \{7, 30\}$. The variable Boolean logic X_6 varies from 3 different logical operators NOT, OR, and AND, and with up to 3 different combinations of all these operators supported in a given query, e.g., $X_6 = \{1, 3\}$. The variable Truncation X_7 varies from 1 Truncation per word to up to 5 Truncations per word, e.g., $X_7 = \{1, 5\}$. The variable Truncation varies from 1 word searched by one of the search engines studied to up to 5 different multi word per search, e.g., $X_8 = \{1, 5\}$. The variable portion of the Web indexed X_9 varies from the title of the Web page to other sections of the Web page including the whole page, e.g., $X_9 = \{.20, 1.0\}$. The second set consists of 69 combinations of the variables X_1, X_2, X_3, X_4, X_5, X_6, X_7, X_8, and X_9 with 6 different set values for each variable. The test set of 209 points was constructed with all possible combinations of 20 equally spaced values of each input variable from 0 to 100. Ten replications of the computer simulation were made for each combination of X_1, X_2, X_3, X_4, X_5, X_6, X_7, X_8, and X_9 for the test set, and both training sets; the mean t_c of each set of ten provided the correct, or target, answer, as was done in ([72]). Models A and B test the ability of approximation techniques to extrapolate beyond the 29 data points used to develop the approximation tool. Models C and D examine the ability of the approximation techniques to interpolate between 69 data points. Two different methods of presenting the data to the approximation tool were used; one used the mean t_c value over the ten replications (Models A and C), and the other used the individual t_c values from each replication (models B and D). These are summarized in table 5.1. Models A and B used full, first-order multiple linear regression models. Models C and D used full, second-order multiple linear regression models. These regression models were selected because they are used typically in response surface methods ([72]). The linear regression models obtained in each case are shown in Table 5.1. The F test for regression was significant for $\alpha = 0.01$ for regression models B, C, and D. Note that regression models A and B are the same model, and that C and D the same model. This is just a manifestation of the regression attempting to estimate the expected value of the output for any particular input value. The benefit of including all of the data in regression models B and D is that lack of fit tests can be performed, although this was not done here.

Table 5.1. Data Presentation Methods Used

Model Label	Data Set	Target Data Presentation Method
A	2^9 points	Mean
B	2^9 points	Individual replications
C	6^9 points	Mean
D	6^9 points	Individual replications

Table 5.2. Regression Models

Model Label	Regression Equation PR = Percentage Rejected	PValue of Statistic	R^2
A	$PR = \sum_{i=1}^{9}(A_i X_i)$	.45	.78
B	$PR = \sum_{i=1}^{9}(A_i X_i)$	$4.3E^{-12}$	.74
C	$PR = \sum_{i=1}^{9}(A_i X_i) + \sum_{i=1,(i<j\,and\,j\neq i)}^{i=9}(B_{ij}X_i X_j)$	$4.1E^{-15}$	.76
D	$PR = \sum_{i=1}^{9}(A_i X_i) + \sum_{i=1,(i<j\,and\,j\neq i)}^{i=9}(B_{ij}X_i X_j)$	$4.3E^{-110}$	.75

The regression in model **A** and **B** of the first order linear equation is of the form:

$$PR(X_1, X_2, X_3, X_4, \cdots, X_9) = \sum_{i=1}^{i=9}(A_i X_i) \tag{5.26}$$

Where PR stands for percentage rejected, A_i are constants that were identified during simulation, X_i represent the 9 input variables. This process is a complex one since different input variables have different range values and different sampling values between these different ranges had to be tested for the best fit among all these constants.

The regression model of the second order linear equation **C** is of the form:

$$PR(X_1, X_2, X_3, X_4, \cdots, X_9) = \sum_{i=1\ (i<j\,and\,j\neq i)}^{i=9}(B_{ij}X_i X_j) + \sum_{i=1}^{i=9}(A_i X_i) \tag{5.27}$$

Where PR stands for percentage rejected, A_i and B_{ij} are constants that were identified during simulation, X_i represent the 9 input variables. If each input variable had 6 different sampling values, there are 10077696 (6^9) training sets possible for the different sets of values of the input parameters considered. Only a few of these different training sets were selected that range from 10000024 points minimum to 69 points maximum. Out of all of these possible training sets, 9000 were investigated. The summary of the results is displayed in table 5.2.

5.3 Comparing the Results of Regression Models and RBF

Figure 5.3 shows the result of the actual output for these different exemplars in Alta Vista on the 9000 simulation cases. Of the 70 different categories of exemplars or queries, 60 were used on training and the remaining exemplars on testing. Figure 5.3 also shows the result of the second order linear regression

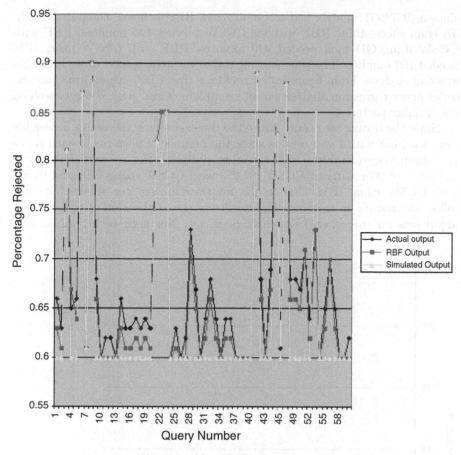

Fig. 5.3. Alta Vista's Performance

of equation 5.27 applied to the different examples. RBF were trained on the 9000 different simulation cases. The model was made out of 9 different input neurons, and 6 output neurons. The reason for the number of output neurons selected is that one output neuron is needed for each range of percentage rejected, e.g. each range spans 5% on the Y Axis of Figure 5.3 and as such 6 neurons are needed for the ranges from 60% to 90%. The number of nodes at the hidden layer depends on the number of exemplars from the 9000 cases for each range. There are more exemplars that cover a given range than other ranges. That makes the neurons learn more about a given range than other ranges. There were 937 nodes at the hidden layer for that particular run.

The success rate was 95.7% on all the examples for RBF without CCSW. RBF with CCSW using GD was 96.7% on all the examples. RBF with CCSW using IPM scored up to 97.6%, which is impressive compared to other studies done with RBF on other benchmark problems. The error was of the order 0.007. The matrix manipulations were heavy at the hidden layer level since a (937*937) matrix had to be inverted. RBF is heavy computationally. To train successfully, RBF without CCSW needed 120 minutes. RBF with CCSW using GD took needed 240 minutes. RBF with CCSW using IPM needed 190 minutes. The time is comparable with the second order linear regression analysis. From figure 5.3 we could see that RBF outperformed second order linear regression analysis on all exemplars. Thus, RBF did very well on the training patterns.

Since the testing set is only limited to ten exemplars, the vector of weights were used but with fewer centers since the number of patterns in each range has shrunk considerably. On the testing patterns, the success rate was of the order of 90% without CCSW, 91.2% with CCSW using GD, and 91.8% with CCSW using IPM. The results for testing were not shown but they follow the results on training. The results of the application of the different algorithms on the other 3 search engines, i.e., Northern Light, yahoo, and

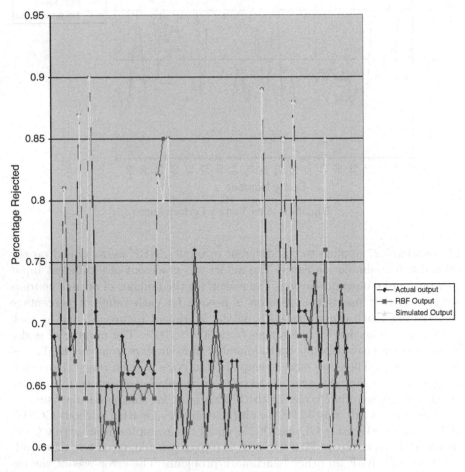

Fig. 5.4. Northern Light's Performance

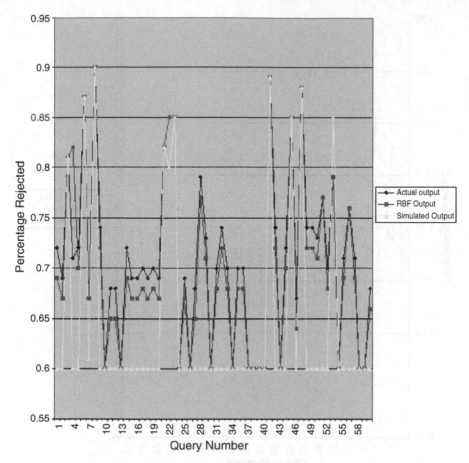

Fig. 5.5. Yahoo's Performance

google in Figures 5.4, 5.5, and 5.6, respectively. Out of all these 4 engines, Alta Vista fared the best in the least amount of rejected pages. The overall number of percentage of rejected pages in Alta Vista is less than the other 3 engines by an average of 3% on all the exemplars. The selected search engines index 26% of existing Web pages on average (estimated around 1 billion Web pages as of 1/1/2000), even 3% less of rejected pages in Alta Vista will mean thousands of less pages that could have been used in the actual search for the queries numbered. Northern Light came next, followed by yahoo, and google. The experimental results of the testing set for both the regression models and the RBF models are provided in Table 5.3. The relatively high mean absolute error (MAE) for both RBF and the regression models indicates the danger in trying to predict responses in regions where training data has not been provided. Models C and D were used to examine the interpolation ability of the approximate methods. The results in Table 5.3 indicate that RBF, regardless of training on mean data or individual data or on individual

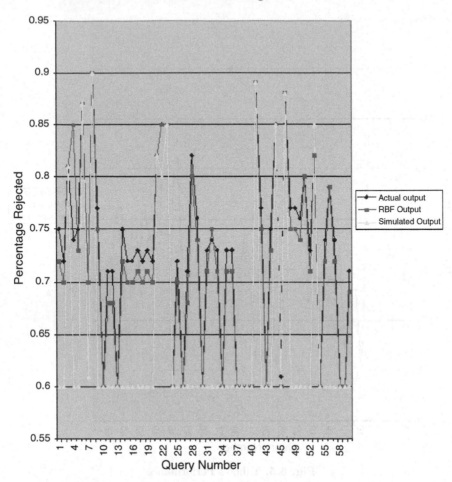

Fig. 5.6. Google's Performance

Table 5.3. Comparison of Test Results (M.A.E. stands for mean absolute error)

Model Labels	Number of Points	Number of training Vectors	Regression M.A.E.	RBF M.A.E.
A	2^9	2^9	3.4	3.2
B	2^9	2^9	4.3	3.7
C	6^9	6^9	1.81	0.98
D	6^9	$10 * 6^9$	1.61	1.31

replications, outperformed the corresponding regression models. These results were significant at $= 0.01$ for a matched t test. This indicates that RBF models can provide better results for interpolating from a computer simulation than the multiple first order and second order regression models commonly used in response surface methods.

The next step was to reconsider the large number of inputs that is needed in the process. Studies suggested testing whether the actual meta-model supports the reduction of the number of inputs without affecting the performance of the output results, especially when the number of inputs is high. This idea was tested effectively in our meta-model. Given the fact that we know the output of RBF with the complete set of 9 inputs, we could eliminate in each run a given input and calculate the output of RBF and compare it with the known output for 9 inputs. If the error between both values was still smaller than a given threshold that particular input could be ignored. Nine different runs were investigated each time eliminating a given input from the calculation. The results show that RBF was successful in eliminating 2 inputs out of 9 (i.e., Response time, and portion of the Web indexed) without affecting the performance of the RBF networks and was still better than second order linear regression analysis with 9 inputs. The fact that these 2 inputs did not affect the performance of RBF neural networks can be interpreted as follows:

1. Response time for the four Web search engines did not vary greatly, and the average waiting time (i.e., between issuing a search command and displaying the first batch of search results on the screen) for every search was in the range of 1-10 seconds. Also, no significant difference was found in Response time between peaks or off hours of Web usage. Some of these queries were tested on the weekends and some during weekdays. Also all queries were tested on a constant interval for a period of one month to account for the Coverage variable in the study.

2. In regard to the portion of the Web indexed, we know that whenever a Web search request is issued, it is the Web index generated by Web robots or spiders, not the Web pages themselves, which has been used for retrieving information. Therefore, the composition of Web indices affects the performance of a Web search engine. One of the main ingredients of the makeup of a Web index is the portion of the Web page indexed (e.g., titles plus the first several lines or the entire Web page). It is well established in the Internet industry that this parameter depends largely on the power and sophistication of the hardware and software that make the Web index or database. In that case it seems that these 4 Web search engines had similar capabilities and as such this variable did not dramatically change the results of the simulation. When trying to eliminate more than 2 inputs at a time the network degraded enormously and the success rate dropped by an average of 50% on the training patterns.

In addition, networks trained on the individual replication output data had better generalization performance than networks trained on just the averages of the simulations output. In other words it is best when approximating stochastic computer simulations to use the "noisy" individual replications rather than the "quiet" average values. The iterative RBF meta-model performed well at approximating computer simulations. Other researchers, for comparison purposes when developing their own RBF meta-model approach, can use the iterative RBF meta-model approach. This contribution is needed in the field of artificial neural networks because there are many different existing and emerging ANN procedures for performing approximation and estimation tasks. Why Alta Vista fare best on the number of rejected Web pages for the broad set of real world queries used in this study? Although Northern Light holds the first place in the percentage of Coverage of Web pages to the size of WWW pages, followed by Alta Vista, google, and yahoo yet it did not outperform Alta Vista. The answer lies not only in the Coverage of each search engine to the estimated size of the WWW, but also other important parameters have to included in the analysis. One of the parameters that may be the key to how well search engines fared in this study is the "age of a new result for a given search" found in the input variable "Coverage". Searches have to be tested over a period of time in order to validate studies. Alta Vista fared the best on the median age of the Coverage followed by Northern Light, yahoo, and then google. That explains why Alta Vista did better than Northern Light in the number of rejected Web pages. Also that explains why yahoo can recover its low Coverage of the WWW with its high Coverage compared to google. This also implies you do not have to increase the size of your database to do better on real-world queries. A formula on how to measure the goodness of a search engine (SE) emerges as follows from this study on the set of queries (SQ) used:

$$G(SE, SQ) = \alpha(Coverage) + \beta(Age) \tag{5.28}$$

where Coverage is the Coverage with the estimated Web size at the time of the experiment, and Age is the "median age" of new document in the result set, α and β are each between 0 and 1.

5.4 Summary

A major area of research is in the experimental design of neural networks as meta-models of computer simulations. This research has filled in a critical need in the designs that take into consideration both the development (training) and the evaluation (testing and validation) of meta-models. This study showed how a meta-model is constructed using a training set for adjusting the weights, the centers and the spreads, and one test set for determining when to stop training and a second test set for evaluation of the generalization ability of the meta-model. The use of an iterative approach rather than baseline ANN meta-modeling reduced simulation runs near 40%. This research investigated the use of extreme values observed from the simulation. Integrating values of the average output for a particular combination of the input parameters that are much larger than all other average output values was possible because of the nature of RBF neural networks. RBF neural networks are de facto neural networks when off line analysis is needed and where speed is not the goal but accuracy is the main concern. It has been shown that on the WWW simulation, RBF performed better than second order linear regression. The number of original inputs in the evaluation of the quality of search engines which was quite high, 9, was reduced during the study to 7, which is major finding compared to other studies which suggested the reduction but failed to achieve it on an actual simulation data. Higher order linear regression analysis techniques could have been used to slightly improve on the results of the second order linear regression analysis but the authors believe that this would have escalated the time to calculate the Simulation experts should use RBF as a model to build on in their actual domain of simulation. Research along these lines is essential to ensure that this tool is properly integrated with other emerging technologies to provide World Wide Web simulation experts' ways to build quality search engines to harness all the Web pages that are available on the Web. The increased complexity of the Web, the almost exponential increase of the number of Web pages since 1995, the distribution of Web pages which follows a power law (*Albert et al.* [3]), and the increased dependency of 100 of millions of users daily on information provided by the Web, make the WWW the most single important information space ever built by Humans. The complexity of the WWW space requires using new intelligent approaches and techniques to explore such a space. Thus, other intelligent techniques to the ones used in this simulation, e.g., rough set theory (*Pawlak* [106]). As rough set theory is being applied successfully in a variety of many real-world problems ranging from medicine to market analysis, applying it to the data mined from the WWW will help better in the ranking of such an "information space". Search engine designers are using a variety of crawling algorithms (agents, bots, and spiders) with a variety of new features not just to compete among themselves, but also to keep up with the exponential explosion of the number of Web pages that are added. New software and techniques are being featured in their technical reports, such as new clustering techniques. These techniques are being applied to better advertise products on the Web to reach a wider audience. New techniques for ranking the results of indexes are being patented to provide services to users where depth is more important than breadth. Web pages are not ranked equally. The notion of equal access is once being marred by the ill biased usage of technology. For instance, the higher the number of links to a Web page makes it more attractive for indexing. Web engine designers use a variety of techniques to choose which page to index. This bias can be understood because the "commodity" is becoming hyperlinks. The richer a Web page is in links, the richer the links to the same Web page make it an ultimate target for indexing. Thus an inherent bias of indexing already exists. If we add to that, the fact that new crawling techniques are adding features such as popularity to guide searching, greater bias in indexing is skewing the results of any query. Google for example, uses the source of incoming links to rank the relevance of pages. President Bill Clinton will have a higher chance of being selected among the pool of Web pages to be indexed than Joe Doe in Arkansas. Instead of indexing being based on content and not popularity, new Web pages will stay "invisible" for a long time from the WWW listings. The increase of the number of studies on the WWW in every journal issue and conference in domains ranging from Artificial Intelligence to Information sciences highlights the importance of the challenges of the WWW. The authors believe more journals and more conferences across disciplines ranging from psychology, art, computer and information science will lead towards a better understanding of the WWW. This effort will ultimately improve the use of the Web by daily users. As more studies on novice and experienced users behavior in searching the WWW are being conducted (*Lazonder et al.* [73]), the authors believe that training students at an early age to use the WWW will become a part of the curriculum that will be integrated within existing multimedia courses in middle schools and high schools. Visual aids will help users better understand their results. Visual aids that will help display not only the set of pages that were explored but also the ones that were left out from their results will help users reformulate their questions to better query the WWW. This constant interaction between the results displayed, the ones left out, and the user needs to be further modeled in a "feedback" mechanism of inhibition and excitement as this already known in any world of discovery. Thus, a "mental model" of the WWW is needed. Discrete structures of trees and graphs that explore the topology of the WWW and the interaction among all the hyperlinks of any given page of the WWW become a hot issue in further studying the WWW. The WWW is rich with information, but which information is indexed in the databases is the issue. When we judge the relevance of an engine, we are judging the relevance of what is being made inclusive in the databases. Thus, this study helped evaluate and estimate the number of Web pages that are left out from a given search result. Providing such information will help both users and designers alike address the issues of accessibility, scalability, limitations, and biases in the design of search engines.

6

Modeling Human Behavior on the Web

6.1 Introduction

Searching for information, in general, is a complex, fuzzy, and an uncertain process. The WWW is a system of storage and retrieval of information characterized by its enormous size, hypermedia structure, and distributed architecture. Traditionally, in paper documents researchers have had clear-cut boundaries to look up information in an index, footnote, etc. On the Web, we are sometimes limited by the novel characteristics of the tool for our search. We must first ask the question of how researchers look at a Web search and do they browse for information. Much research has been done on the process of browsing for information. In browsing, one may not have a specific idea of where the target of the search might be, but have a strategic search pattern in mind which is goal oriented. We search for new information, and we anticipate that the information we want will be available if we follow a certain pathway to it. *Canter, et al.* ([20]) has suggested that there are five different browsing strategies:

1. Scanning: Covering a large area without depth (Breadth Search);
2. Browsing: following a path until a goal is reached;
3. Searching: Striving to find an explicit goal;
4. Exploring: Finding out the extent of information given;
5. Wandering: Purposeless and unstructured globetrotting.

In looking for information, the user is primarily concerned with browsing and searching (though also with exploring to a certain extent that is outside our scope). It is obvious that using hyperlinks gives the user new freedom to browse through documents, but the user must impose a structure upon that browsing. The structure that a user imposes is often a neurological one, a cognitive map or model. Many researchers (*Edwards & Harmon*, [31]; *Hammond* [45]; *choo et al.* [22]) have demonstrated the stressfulness of the transition between the bordered world of paper text and the unbounded world of the Web. *Gedge* ([39]) maintains that the use of any computer interface involves the

G. Meghabghab and A. Kandel: *Search Engines, Link Analysis, and User's Web Behavior*, Studies in Computational Intelligence (SCI) **99**, 105–215 (2008)
www.springerlink.com   © Springer-Verlag Berlin Heidelberg 2008

development of a cognitive model representing the structural and procedural properties of that particular system. In addition to our work in finding the information, we must also have an adequate understanding on how to use the information retrieval system (IRS). More saliently, *Wright* ([130]) has identified four steps that any model for user cognitive maps will have to consider:

1. The mechanics of the system (understanding what button performs what task);
2. Establishing the size and complexity of the information available;
3. Determining where one is in the system and maintaining a sense of that position;
4. Remembering what screens were already visited (and thereby saving extraneous steps).

We know that the human brain's solves problems by computing in a drastically different way than the computer. In studying cognitive mapping techniques we must take into consideration that the brain is complex, nonlinear, and capable of multitask parallel functions (*Haykin* [51]). The brain is capable, in particular, of setting up a neurological mapped network for visual-spatial tasks (such as Web searches) on computers that would take a fraction of time ([51] suggests 1-200 microseconds) that the same task would take on a computer. Cognitive mapping theory focus on the ways an individual navigates through familiar and unfamiliar spaces, such as finding one's way to a store by driving a car. We learn the correct path, and over time that path becomes rooted in a neurological pathway that we find our way to the familiar location automatically (*Streeter and Vitello*, [126]). Since using any computer interface involves visual and spatial skills similar to navigating in a real world space (i.e., finding the goal by following a pattern designed to allow us to hit the targeted goal), then in order to be successful browsers, we must form cognitive maps not only of the structure of cyberspace, but also of how we might best search for that information. All learning research suggests that learning is a direct product of experience. Experience is built around:

1. direct interaction with the material learned; and
2. direct observation of users performing successfully the task.

As we gain experience, making inter-neuron connections of varying strengths (*Hurrion* [58]) form these neural circuits or cognitive maps. In order to master a task, we must then make a synaptically weighted inter-neuron connection that facilitates the task at hand. In the cognitive interface between Web search engines and users, then, we must consider both the Web search engine characteristics and the characteristics of those users, that is, the cognitive maps they construct of their searches using the search engines. We propose that the experience that newcomers to the Web create these maps can best be understood by creating a fuzzy cognitive map and compare it with a markovian modeling already established by *Meghabghab* ([85]). Understanding the user's cognitive map in this way will allow educators to design instructional programs that will minimize extraneous attempts to search for information and thereby improve the transition from paper to the Web.

6.2 An Experimentation of User's Web Behavior

A quantitative method was employed to capture Web activities of the participants in this study. Even though many factors differentiate between different users, a model has to be general enough to capture the common behavior among all users. The purpose of this section is to record in a model how a user navigates through the search engine and allows one to derive behavior metrics such as how many times a certain browsing function is invoked during a session or the average length of a browsing session.

6.2.1 User Cognitive Model of a Browsing Session

Consider a search engine in which users can perform the following functions:

1. Connect to the "Home page" of the search engine. Users can loop over the Home search engine.
2. Search the Web for answers to query after using keyword search, boolean logic to combine keywords searches. A user is in that stage after being given a query and the process of problem solving has started. Users can also loop over a search they have done already.
3. Depending upon the set of results returned by the search engine, the user can just be evaluating the returned set especially, in a domain where he/she is unfamiliar with it. This constitutes a scanning state. This state constitutes a "shallow interpretation" of the set of results returned. It is a part of the "browse" state. This state constitutes a final step in knowledge filtering.
4. Browse the links provided in the resulting search set to the actual query. This state will replace both implicit states of scanning or exploring. Users can loop over a Web page that they already browsed.
5. Select one of the links that resulted from the search and view additional information such a brief description, summary, ranking, etc. If the extent of information given is satisfactory and relevant and no browsing is needed, then the user is into a select stage. This stage is a consequence of the fact that users have a goal in mind and they are in the final decision state, which matches their knowledge comprehension state. Users can also loop over a Web link that they have already selected.
6. Backtrack to previous screens if the link viewed does not contain answer. Basically backtracking constitutes moving between different stages of browse and select, or between select and search, or between search and home.

7. Users do not wander on the Web aimlessly when they are given a query. They have a predefined task, which is to query the Web. The goal of the task is to find an answer for the query. They either "succeed" or "fail" in the task. They may loop over the same Web page, browse the same Web link, select over and over the same link but not aimlessly.

These different stages or states constitute the framework under which we will build our cognitive map of the WWW. State S_i symbolizes a state at time i. There are different constraints that users go through. These are Time constraints and Information constraints or overload. Time constraints are times imposed to answer a given query on the WWW. Information overload results from a search that returned lots of information and a refined search is needed. Also, a cognitive constraint can be added, which states that users cannot have two behaviors at any given time. If they are querying the Web they cannot be browsing at the same time. They cannot be exploring the search result and exiting the search engine at the same time. They are in a redefined stage of thinking on the Web at any given time. These stages are unique and cannot be violated at any time. For any time i different than j, state S_i is different from state S_j, which can be stated as follows:

$$\forall\ i \neq j\ :\quad S_i \neq S_j \tag{6.1}$$

Figure 6.1 can help deduce a matrix of constraints between these different states. Table 6.1 is a double entry table of the edges between different states.

The general equations, which relate the nodes to the edges, in Figure 6.1 of the "Markovian Model" of the User Cognitive Map are:

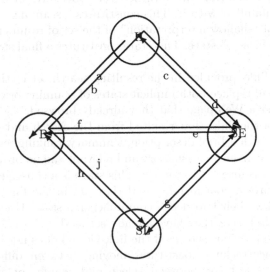

Fig. 6.1. Markovian Model of the User cognitive map in a browsing session

Table 6.1. Edges between the different states

	Home	Search	Browse	Select
Home	e_{HH}	e_{HSe}	e_{HB}	0
Search	e_{SeH}	e_{SeSe}	e_{SeB}	e_{SeSl}
Browse	e_{BH}	e_{BSe}	e_{BB}	e_{BSl}
Select	0	e_{SlSe}	e_{SlB}	e_{SlSl}

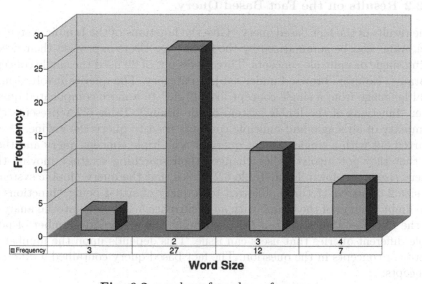

	1	2	3	4
Frequency	3	27	12	7

Word Size

Fig. 6.2. number of words vs. frequency

$$V_H = 1(\text{where H = Home}) \tag{6.2}$$

$$V_s = \sum_{k=1}^{k=N} V_k e_{ks} \tag{6.3}$$

where N is the number of states in the model. Applying equations 6.2 and 6.3 to Figure 6.2 and using Table 6.1 yields:

$$V_{Se} = V_H e_H S_e + V_B e_B S_e + V_{Sl} e_{Sl} S_e + V_{Se} e_{SeSe} \tag{6.4}$$
$$V_{VB} = V_{Se} e_{Se} B + V_{Sl} e_{Sl} B + V_B e_{BB} + V_H e_{HB} \tag{6.5}$$
$$V_{Sl} = V_B e_{BSl} + V_{Sl} e_{SlSl} + V_{Se} e_{SeSl} \tag{6.6}$$
$$V_H = V_B e_{HB} + V_{Se} e_{SeH} + V_H e_{HH} \tag{6.7}$$

Rewriting equations 6.4 to 6.7 and replacing the arcs between states with the values a, b, c, d, e, f, g, h, I, j, k, l, m, and n yields:

$$H = bB + dSe + nH \tag{6.8}$$

$$B = gSl + eSe + lB + aH \tag{6.9}$$
$$Se = cH + fB + kSe + jSl \tag{6.10}$$
$$Sl = mSl + hB + iSe \tag{6.11}$$

The values of the different edges have to be calculated by users that will effectively browse the Web. The next section will describe the process in detail.

6.2.2 Results on the Fact Based Query

The results of the fact based query "Give the functions of the Limbic system of the brain" can be summarized by the followings: Users expressed their needs using single or multiple concepts. Three users out of 20 used the single concept "Brain". No users browsed under subject category. The queries involved natural language from a single concept like "brain" to a more complicated query like: "functions of the limbic system of the brain?" Table 6.2 represents the summary of all single and multiple concepts used to query the engine. Users started out with a single concept query or a multiple concept query and then as that they got involved they changed their searching strategy moving the query size up or down. From Table 6.2 we see that the query "limbic system" of size 2 occurred 17 times, followed by a query of size 4 being "functions of the limbic system of the brain" with an occurrence 5. A combinatorial analysis of the fact-based question can help determine the maximum number of possible different queries that users can issue. This depends upon the number of words or concepts in the question. The fact based query contained 4 different concepts:

Table 6.2. Occurrence of queries for the fact based query

Queries Used	Occurence	Percentage
Brain Limbic System	3	6.02
Brain Limbic System Functions	2	4.08
Brain Functions	1	2.04
Limbic System of the Brain	5	10.2
Limbic System	17	34.69
Functions of Limbic System	3	6.12
Limbic Brain	1	2.04
Brain System	2	4.08
Brain Function System	1	2.04
Parts of The Brain	1	2.04
Systems of the Brain	1	2.04
Brain	3	6.12
"Limbic" + "Brain"	1	2.04
Human Brain	3	6.12
Functions of the Limbic System of the Brain	5	10.2
Total	49	100

- If we use 1 word out of the set of 4 words, then the number of combinations is C(4, 1) = 4.
- If we use 2 words out of the set of 4 words: C(4, 2) = 6.
- If we use 3-word out of the set of 4 words: C(4, 3) = 4.
- If we use 4-word out of the set of 4 words: C(4, 4) = 1.

It happens that users in the case of a 2-word concept search they used 4 combinations out of possible 6 with query "Brain System" and "Systems of the Brain" being considered one and the same. In the case of a 3-word concept search, users used 3 out of 4 possible combinations with "Brain Limbic System" and "Limbic Systems of the Brain" being one and the same. In the case of 4-word concept search, users used 2 combinations out of a set of 4 possible combinations. The total number of possible combinations of a 4-word concept query is:

$$\sum_{i=1}^{i=4} C(n,i) = C(4,1) + C(4,2) + C(4,3) + C(4,4) = 15 \qquad (6.12)$$

That explains the extent in table 6.2 of the number of possible queries used. Users used 10 different queries out of all possible 15 combinations with one being repeated twice in our study to show the use of "" in the query: "limbic"+"brain". Also users used 2 queries with concepts that were not there in the original query: Parts of the brain and Human Brain. The first query reflects user's knowledge that the limbic system is a part of the brain. The other query added the qualifier that the brain belongs to humans. One user misspelled limbic into limbric. He/she combined the misspelled word with other concepts in the original question. Table 6.3 summarizes the result of the user with all possible queries used.

Users never used terms that were not related to the question asked. They used queries that were related but failed sometimes to understand what they read. They looped back and forth between different Web pages showing frustration during traversal. This happened for users who were successful and users who failed in their task. The search strategies students employed impacted the success/failure of information retrieval.

Figure 6.2 shows the relationship between the number of words used in a query and the frequency of its usage. It shows that users used words of size 2 27 times compared to words of size 3 12 times and words of size 4 7 times

Table 6.3. Misspelled queries used and their frequency

Queries Used	Frequency
Limbric Brain System	2
Limbric System	3
Limbric Brain	2

Table 6.4. Occurrence of Queries as a function of the length of the fact-based query

Queries Used	Frequency	Percentage
Size 1	3	6.12
Size 2	27	55.1
Size 3	12	24.49
Size 4	7	14.29
Total	49	100

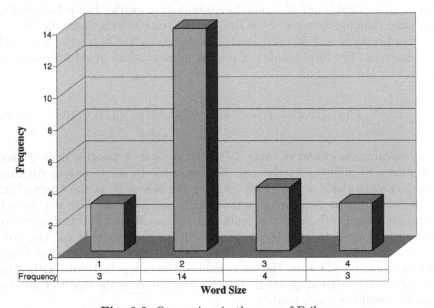

Fig. 6.3. Query sizes in the case of Failure

(see table 6.4). They were more comfortable with words of size 2 than words of size 4. How many users who used words of size 2 were successful and how many failed? Figure 6.3 displays the size in words on a failure. Users who failed to find an answer used no words of size 1 in their searches. They used mostly words of size 2 13 times. They also used words of size 3 8 times out of 9 on the total amount of words of size 3.

Figure 6.4 displays the number of times users used queries of different sizes and were successful. Users who succeeded used equal number of words of size 2 than those who failed, but used words of size 3 less than those who failed, and equal in number of word size 4 to those who failed. What percentage of users changed the length of concepts of their query during a session as they refine their query strategy?

Table 6.5 displays the results on failure. Table 6.6 displays the same results on success. It is obvious that users who were successful did not change the word size of their query most of the times.

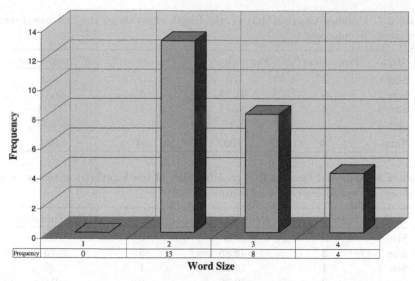

Fig. 6.4. Query sizes in the case of Success

Table 6.5. Change in word length vs.the number of users that failed

Word Length	Number of Users
Up	2
Down	1
Did Not Change	3
Up → Down →	1
Up → Down	
Up → Up → Down	1

Table 6.6. Change in word length vs. the number of users that succeeded

Word Length	Number of Users
Up	1
Down	1
Did Not Change	8
Up → Down	2

This does not mean they did not do many searches of the same word size. On the other hand, users who failed changed their query strategy most of the time. It is difficult to use the average number of searches as an indicator of success or failure on the fact-based query, especially that these averages are very close: average on success is 2.75 and average on failure is: 3.13. But is the change of word length during a session as users refine their query strategy an indication of success or failure? Users repeated some times the same query in different phases of their search strategy. Users also modified some times

Table 6.7. Modified Queries(MQ) vs. the length of terms on the fact-based query (F=Failure, S=Success)

MQL	Frequency(F)	Percentage(F)	Frequency(S)	Percentage(S)
Size 1	0	0.0	0	0.0
Size 2	9	64.29	6	66.67
Size 3	4	28.57	1	11.11
Size 4	1	7.14	2	22.22
Total	14	100	94	100

Table 6.8. Repeated Queries(RQ) as a function of the length(L) of terms on the fact-based query (S=Success, F=Failure)

RQL	Frequency(F)	Percentage(F)	Frequency(S)	Percentage(S)
Size 1	0	0.0	0	0.0
Size 2	3	42.86	3	75
Size 3	1	14.29	1	25
Size 4	3	42.86	0	0.0
Total	7	100.00	4	100

their queries at different phases of their search strategy. Some of the modified queries were of the same length and some of the modified queries changed in length. Table 6.7 shows the results of the modified queries in success and failure. Users who succeeded modified their queries 39.1%, while those who failed modified their queries 60.9%. It is also noticeable that users never modified their queries to length 1 in both success and failure cases. This explains the difference in the number of queries between successful users and unsuccessful users. Table 6.8 shows the results of the repeated queries in success and failure. Users who succeeded repeated their queries 36%, while those who failed repeated their queries 64%. It is also noticeable that users never repeated their queries of length 1 in both cases of success and failure. This explains the difference in the number of queries between successful users and unsuccessful users. From Table 6.4 we conclude that queries of length 2 or more represent 85.71% of the total number of queries. While *Spink et al.* ([125]) found most of the queries per session on Excite are of length 2 or less on a 2.5 million samples, our study with only 20 users cannot constitute an absolute reference for other studies.

From Figure 6.5 we conclude that our users visited on average less than 7 Web pages, spent less than 7 minutes to answer the query, used less than 3 Web searches to find their answer, and took them less than 13 moves between all the different states to get to the target answer. Figure 6.6 displays the final results in terms of average number of searches on failure, the average number of searches on success, the average number of Web pages visited on failure and the average number of Web pages visited on success. Also it shows the average time used on failure and the average time on success. It is important

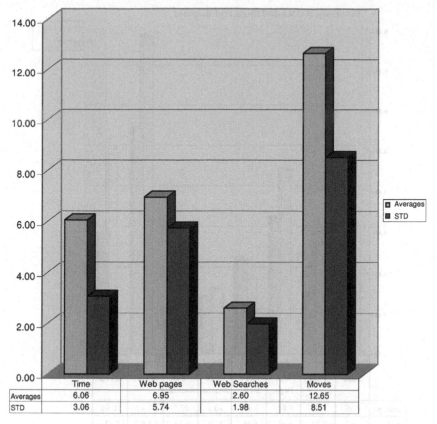

	Time	Web pages	Web Searches	Moves
Averages	6.06	6.95	2.60	12.65
STD	3.06	5.74	1.98	8.51

Fig. 6.5. Statistics of All Summary data

to note that any Web page that a user used was counted as Web page that is visited. Some Web pages may have been visited more than once and we still counted them every time they were used. The researchers of the study feel that the high number of Web pages visited by users who failed is a solid discriminating factor in this study of the cognitive behavior of our users. Users who failed viewed 3 times more Web pages those who succeeded on average. On average, users who failed were not able to evaluate the information retrieved and many times looped over the same Web page without being able to judge the relevancy of the information retrieved for the given query.

Another important measure is the number of moves done by users. Any move from any state to any other state is counted a move. Counting the number of moves for a given session is an important value needed for our equations. We separated these moves between users who failed and those who were successful. Figure 6.7 shows a user's cognitive map. The user failed to provide the correct answer to the query for example. The user had a session of 34 moves around the different states.

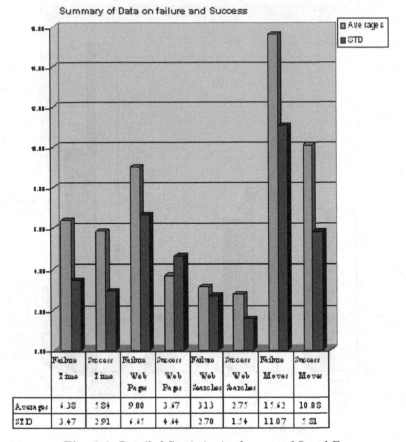

Summary of Data on failure and Success

	Failure Time	Success Time	Failure Web Page	Success Web Page	Failure Web Searches	Success Web Searches	Failure Moves	Success Moves
Averages	6.38	5.84	9.00	3.67	3.13	2.75	15.62	10.08
STD	3.47	2.91	6.65	4.64	2.70	1.54	11.07	5.81

Fig. 6.6. Detailed Statistics in the case of S and F

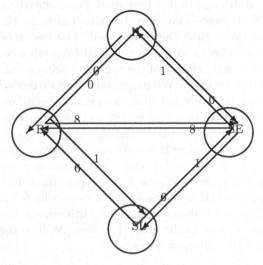

Fig. 6.7. A User Cognitive Map showing all the moves in case of a failure

We could say that the transition probability out of Home to Search is: 1/34, the transition probability out of Search to Browse is 8/34, the transition probability out of Browse to Search is 8/34, the transition probability out of Search to Select is 1/34, the transition probability out of Select to Search is: 1/34, the transition probability out of Browse to Select is: 6/34, the transition probability out of Select to Browse is: 6/34, the transition probability out of Select and looping back to itself is 1/34, the transition probability out of Search and looping back to itself is 1/34, the transition probability out of Browse and looping back to itself is 0/34, the transition probability out of Home to Browse is 0/34, the transition probability out of Browse to Home is: 3/34. We will calculate the average of all the states on success and on failures well as the average of all the transition probabilities on failure and on success. Also, the standard deviation on all states and transition probability on success and failure is calculated. Applying equations 6.8 to 6.11 yield the following values for the states B, Se and Sl: B = 35.97, Se = 45.22, Sl = 51.93. Users in general selected Web in the pages more than they searched, searched more than they browsed, browsed more than they stayed home on a fact-based query. Applying equations 6.8 to 6.11 with the actual data are displayed in Figure 6.8. The equations yield the following values for the states B, Sl and Se: H = 1, B = 36.25, Se = 3.86, Sl = 7.31. Thus, the characteristics of the users who succeeded on a fact-based query: they searched more than they selected Web pages, selected Web pages more than they browsed, and browsed more than stayed Home. In the case of the users who failed the query, equations 6.8 to 6.11 are applied with the actual data. The equations 6.8 to 6.11 yield the following values for the states B, Sl and Se: H = 1, Se = 7.12, Sl = 26.7, B = 17.01. Thus, the characteristics of the users who

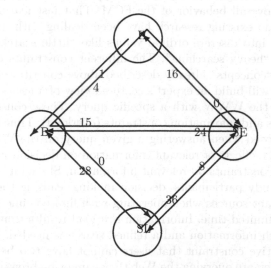

Fig. 6.8. A User Cognitive Map showing all the moves in case of success

failed to answer the query correctly: they selected more than they browsed, they browsed more than they searched, and they searched more than they stayed home.

6.3 Fuzzy Sets Modeling of User's Web Behavior: A New Model

In this section, we propose a fuzzy cognitive map (FCM) of a user's behavior on the Web (Meghabghab [85]). It represents the opinions of experts on how users surf the Web. Experts are divided on what causes users to fail their queries on the Web or what keep them in being successful regardless of the domain or the task at hand. This section explores that a viable FCM model can be developed and some limit-cycle equilibrias (see appendix) are uncovered. A FCM limit cycle repeats a sequence of events and actions. Limit cycles can reveal cognitive and behavioral patterns of users on the Web. An adaptive FCM is built around correlation encoding and differential hebbian learning to reflect the map behavior in time. The dynamics of the adaptive hebbian learning FCM built show for the first time, since no other studies have shown that, users that are search oriented and unsuccessful can learn to become successful. Also, the dynamics of the adaptive hebbian learning FCM built show for the first time that users that are browse oriented and unsuccessful can learn to become successful. The FCM build shows how successful users can stay successful with new uncovered rules on learning. The behavior of the FCM can also get stuck in a minimum and chaotic behavior is detected. Few FCMs implementation in the literature have shown such an erratic behavior. Also sub-concepts can be made to appear on searching and browsing to improve the overall behavior of the FCM. That last feature needs to be exploited since no existing research have been dealing with dividing searching or browsing into discrete ordinal values like "little searching", "average searching" and "heavy searching". The different constraints combined with the "states" or "concepts" already described above constitute the framework under which we will build an expert's centered view of a user's cognitive map while searching the WWW with a specific query. These constraints include Time constraints and Information constraints (overload). Time constraints are time limits imposed when answering a given query on the WWW. *Higgings* [56] examined the effects of relevant information on decision-making and the effects of Time constraints on relevant information. She found that time limits influenced study participants' decision making, causing them to rely less heavily on relevant sources when operating with limited time than when operating with unlimited time. Information overload results from a search that returns too much information and a refined search is needed. Added to this, there is a cognitive constraint that users cannot have two behaviors at any given time. If they are querying the Web they cannot be browsing at the same time. They cannot be exploring the search result and exiting the search engine

at the same time. The following concepts summarize the above and constitute the building blocks of a cognitive map of users searching or browsing for
the information on the WWW with a specific query, and the constraints that
are imposed on them or self imposed (*Gross* [42]) which lead to their success or failure: C_1 =Search, C_2 =Browse, C_3 =Time constraints (T.C.), C_4 =
Information constraints (I.C.), C_5 =Success, C_6 =Relevance, C_7 = *Failure*.

6.3.1 Review of Literature on FCM

Fuzzy cognitive maps are fuzzy digraphs that model causal flow between concepts (*Kosko* [67, 68]). They are dynamical systems that relate rules. They
have been used to model virtual worlds (*Diskerson & Kosko* [29]), artificial life
(*Dickerson & Kosko*, [30]), and vehicle behavior in traffic simulations (*Hong
& Dickerson* [57]). FCMs have been used for modeling systems that have
uncertain and incomplete models that cannot be easily expressed as equations. Some examples are modeling human psychology (*Craiger & Coovert*
[26]; *Hagiwara* [44]), modeling slurry rheology (*Banini & Bearman* [7]), and
on-line fault diagnosis at power plants (*Lee et al.* [75]), societal issues such
as the politics of racism in South Africa (*Taber* [127]). All of these problems
have some common features. The first is the lack of quantitative information
on how different variables interact with one another. The second is that the
direction of causality is at least partly known and can be articulated by a
domain expert. The third is that they link concepts from different domains
together using arrows of causality. These features are shared by the problem
of modeling a user's searching behavior on the Web in which knowledge is
available at different levels of detail. The FCM can take this partial information and structure it. The FCM will allow the users to explore the implications
of different data paths and search for critical paths using the known information on the direction of causality, but without making assumptions about the
size of the effect. An FCM is a digraph structured as a collection of nodes
and arcs. Nodes, called concepts, are system variables; their values change
over time. Connections among concepts, the arcs or edges of the graph represent causality. That is, a weighted directed edge that connects concept C_0
to C_1, pointing from C_0 to C_1, tells that C_0 causes C_1. An edge may connect
any concept to any other concept. An edge may also connect a concept to
itself, indicating that the future value of the concepts depends on the concept's current value. Each concept, or node, has a strength value traditionally
ranging from 0 (no strength) to 1 (full strength). The 0 to 1 range is arbitrary, and some users prefer bipolar values ranging from -1 for maximum
negative representation, through 0, for ambivalence, to +1, for maximum positive representation. The runtime operation of an FCM consists of calculating
the next value of each concept in the FCM from the current concept and
edge values. An FCM is a sampled-data system in which the collection of
current concept values represents the current overall system-concept status.
At each time increment, the next value of each dependent system concept is

calculated from the current concept and edge values. Several techniques are available for calculating concept values of a system. The most common one is a normalized sum of products. The first step is to take the product of each source concept value and connecting edge value, and then sum these products and normalize, or "squash," the result into the range of allowable concept values. The purpose is to cause sums greater than one, which can occur when several products are summed, to be monotonically mapped into the 1 (or 0 to 1) range. Appropriate normalizing functions are those based on the sigmoid function popular in neural networks, linear functions of varying slopes, and a number of probability density functions (*Haykin* [51]). The links in the FCM can be redefined as continuously varying functions to reflect this information. At time t, the state of the FCM is the concept vector $C = (C_1, C_2, \cdots, C_n) \in [0,1]^n$ or a point in the fuzzy-cube state space. The n concepts all belong to the state C to some degree at time t. The simplest FCM acts as an asymmetrical network and converges to limit cycles. An FCM carves the fuzzy-cube state space into k many attractor regions (see appendix) or regions of great stability. Some regions contain fixed points (see appendix) or limit cycles (see appendix) or other regions contain attractor chaotic (see appendix). FCM are temporal associative memories (TAM) (*Dickerson & Kosko* [29]) that encode causal relationships. The connections in an FCM can be learned using neural learning laws (*Kosko*, [67, 68]) or by constructing fuzzy sets using user determined thresholds. The Deterministic differential hebbian learning law ([67]) correlates changes in causal concepts as well as the presence or absence of a concept. Hebbian learning is a postulate of *Hebb* ([52]) which specifies: "When an axon of cell A is near enough to excite a cell B and repeatedly takes part in firing it, some growth or metabolic changes take place in one or both cells such that A's efficiency as one of the cells firing B, is increased". The weight matrix updates only when a causal change occurs at the input. Otherwise most causal inferences will be forgotten as the signal exponentially decays. This works well for political or economic problems where many different experts hold a variety of views (*Taber* [127]). The weights may be found by surveying experts in the field or by neural learning. Threshold FCM quickly converges to stable limit cycles or fixed points. These limit cycles show "hidden patterns" in the causal Web of the FCM.

6.3.2 Applying FCM to User's Web Behavior

As stated above, seven concepts are chosen as building blocks of a cognitive map of users searching or browsing for information on the WWW with a specific query, and the constraints that are imposed on them or self imposed ([42]) which lead to their success or failure: C_1 = Search, C_2 = Browse, C_3 = Time constraints (T.C.), C_4 = Information constraints (I.C.), C_5 = Success, C_6 = Relevance, C_7 = Failure. Most causal learning schemes are unsupervised in nature since no one knows the "real" causal structure of the world. Experts are divided on how failure can affect all the concepts involved in the study. For example some view a positive influence of failure on concepts C_1, C_2, C_3, and C_4, while some view them as having negative influence on the same concepts. Thus in matrix E below, one can see that:

$$\mathbf{E} = \begin{array}{c} \\ C_7 \end{array} \begin{array}{ccccccc} C_1 & C_2 & C_3 & C_4 & C_5 & C_6 & C_7 \\ \left[\pm 1 & \pm 1 & \pm 1 & \pm 1 & 0 & 0 & 0 \right] \end{array}$$

This says that failure depends on too much search or too little search. Failure also depends on too much browse or too little browse. Too much information can cause failure and too little information can cause failure. Too little of time to find information can cause failure and too much time to find information can do exactly the same. To build the values of the below matrix E, experts in the area of Web browsing were asked to tell for example whether "browsing" has a positive influence on success of answering a query, negative influence on success of answering a query, or no influence at all. The matrix E below represents their views.

6.3.3 The Case of Positive Influence of C_7 on the Other Concepts

The below matrix E that represents the summary of all responses gotten from experts that were selected to identify the influences positive, negative, or none among the 7 concepts. Figure 6.9 visualizes the influences among the concepts. (only the negative influences is shown). For example, it says that C_1 has no influence (no arrow) over C_1, has negative influence over C_2, and positive influence over C_4 (which is not drawn for readability purpose).

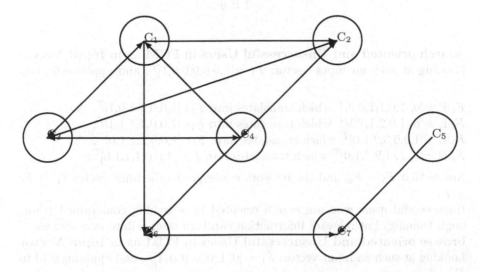

Fig. 6.9. FCM of User's experience in the case of positive influence of C_7

$$E = \begin{array}{c} \\ C_1 \\ C_2 \\ C_3 \\ C_4 \\ C_5 \\ C_6 \\ C_7 \end{array} \begin{array}{ccccccc} C_1 & C_2 & C_3 & C_4 & C_5 & C_6 & C_7 \\ \left[\begin{array}{ccccccc} 0 & -1 & -1 & 1 & 1 & -1 & 1 \\ 1 & 0 & -1 & -1 & 1 & -1 & 1 \\ -1 & -1 & 0 & -1 & 0 & 1 & 1 \\ -1 & -1 & -1 & 0 & 0 & 0 & 0 \\ 1 & 1 & 1 & 1 & 0 & 1 & -1 \\ 1 & 1 & 0 & 1 & 1 & 0 & -1 \\ 1 & 1 & 1 & 1 & 0 & 0 & 0 \end{array}\right] \end{array}$$

FCM's dynamics depend on the dynamics of the concepts nodes C_i and causal edges. The edges e_{ki} are constant weights with values of -1, 0, or 1 and only the nodes change in time. Thus the value of a concept node C_i at time t_{n+1} is a function of the bounded sum S of all the C_{ik} that influence C_i at time t_n:

$$C_i(t_{n+1}) = S(\sum_{k=1}^{N}(e_{ki}(t_n))C_k(t_n)) \tag{6.13}$$

Where S(x) is a bounded signal function. For simple FCM's the sigmoid function:

$$S(y) = \frac{1}{(1 + \exp^{-c(y-T)})} \tag{6.14}$$

If c is large S(y) becomes a binary threshold function:

$$S(y) = 0 \text{ if } y \leq 0 \tag{6.15}$$
$$= 1 \text{ if } y > 0$$

search oriented and Unsuccessful Users in FCM as in Input Vector
Looking at such an input vector: $F_1 = [1,0,0,0,0,0,1]^T$, and applying 6.13 to F_1:

$F_1.E = [0,-1,0,1,1,0,2]^T$ which translates into $F_2 = [1,0,0,1,1,0,1]^T$,
$F_2.E = [0,-1,0,2,1,1,1]^T$ which translates into $F_3 = [1,0,0,1,1,1,1]^T$,
$F_3.E = [1,0,0,3,2,1,0]^T$ which translates into: $F_4 = [1,0,0,1,1,1,0]^T$,
$F_4.E = [0,-1,-1,2,2,1,0]^T$ which translates into $F_5 = [1,0,0,1,1,1,0]^T$.

Notice that $F_5 = F_4$, and the network converges to the limit cycle: $F_4 \rightarrow F_5 = F_4$.

Unsuccessful users who are search oriented by searching constrained (thorough training) but relevant information can turn their failure into success.

browse oriented and Unsuccessful Users in FCM as in Input Vector
Looking at such an input vector: $F_1 = [0,1,0,0,0,0,1]^T$, and applying 6.13 to F_1:

$F_1.E = [1,0,0,-1,1,0,2]^T$ which translates into $F_2 = [1,1,0,0,1,0,1]^T$,

$F_2.E = [2,0,0,1,2,0,2]^T$ which translates into $F_3 = [1,1,0,1,1,0,1]^T$,
$F_3.E = [1,-1,-1,1,2,0,2]^T$ which translates into: $F_4 = [1,1,0,1,1,0,1]^T$.

Notice that $F_4 = F_3$, and the network converges to the limit cycle: $F_4 \rightarrow F_4$. Unsuccessful users who are browse oriented by searching or browsing constrained (through training) information can be successful or can fail, which is a contradiction.

search oriented and successful Users in FCM as in Input Vector
Looking at such an input vector: $S_1 = [1000100]^T$, then applying 6.13 to S_1:

$S_1.E = [0,-1,0,1,1,1,1]^T$ which translates into $S_3 = [1,0,0,1,1,1,1]^T$,
$S_2.E = [1,0,0,3,2,1,0]^T$ which translates into $S_3 = [1,0,0,1,1,1,0]^T$,
$S_3.E = [0,-1,-1,2,2,1,0]^T$ which translates into $S_4 = [1,0,0,1,1,1,0]^T$.

Notice how $S_4 = S_3$ and the network converges to the limit cycle $S_3 \rightarrow S_4$.
successful and search oriented users by searching constrained information (through training) but relevant information can stay successful.

browse oriented and successful Users in FCM as in Input Vector:
Looking at the above as an input vector: $S_1 = [0,1,0,0,1,0,0]$, then applying 6.13 to S_1:
(forcing user to stay successful):

$S_1.E = [1,0,0,-1,1,1,1]^T$ which translates into $S_2 = [1,1,0,0,1,1,1]^T$,
$S_2.E = [3,1,0,2,3,0,1]^T$ which translates into $S_3 = [1,1,0,1,1,0,1]^T$,
$S_3.E = [1,-1,-1,1,2,0,2]^T$ which translates into $S_4 = [1,1,0,1,1,0,1]^T$.

Notice that $S_4 = S_3$, and the network converges to the limit cycle: $S_3 \rightarrow S_4 = S_3$.
successful users who are browse oriented by searching and browsing constrained but relevant information can stay successful or fail, which is a contradiction by itself. The length of the limit cycle of 1 is less than the number of concepts otherwise crosstalk can occur.

Adaptive Fuzzy Cognitive Maps through Correlation of User's Web Behavior

How does a user's behavior change with time while searching the Web? Does a user learn new behavior or reinforce old ones? Is it possible to force a limit cycle or a chaotic behavior on the system? To answer such a question an adaptive FCM is built that uses correlation or Hebbian learning to encode some limit cycles in the FCMs or temporal associative memories (TAM) ([30]). This method can store only few patterns. differential hebbian learning ([68]) encodes changes in a concept. To encode binary limit cycles in matrix E the TAM method sums the weighted correlation matrices between successive states. To encode a cycle, a concept Ci has to be converted to a concept Xi by changing the 0's into -1. Then E is a weighted sum:

$$E = X_1^T X_2 + X_2^T X_3 + X_3^T X_4 + \cdots + X_n^T X_1 \qquad (6.16)$$

The bipolar correlation matrix $X_k^T X_{k+1}$ forms the associative link from state C(k) to state C(k + 1) where X_k replaces the 0's in binary state vector C(k) with -1. Then 6.16 applied to E yields a new E:

$$E = \begin{array}{c} \\ C_1 \\ C_2 \\ C_3 \\ C_4 \\ C_5 \\ C_6 \\ C_7 \end{array} \begin{array}{ccccccc} C_1 & C_2 & C_3 & C_4 & C_5 & C_6 & C_7 \\ \left[\begin{array}{ccccccc} -1 & 1 & -1 & 1 & 3 & -1 & 1 \\ 1 & 3 & 1 & 5 & 3 & -3 & -1 \\ -1 & -1 & 0 & -1 & 0 & 1 & 1 \\ 3 & 1 & -1 & 3 & 5 & -5 & 1 \\ 1 & -1 & 1 & -3 & -1 & 1 & 3 \\ 1 & 1 & -1 & -1 & 1 & -1 & -3 \\ -3 & -5 & -3 & -7 & -1 & 1 & 3 \end{array}\right] \end{array}$$

search oriented and Unsuccessful Users in CE as in Input Vector

Looking at such an input vector: $F_1 = [1, 0, 0, 0, 0, 0, 1]^T$ then applying 6.13 to F_1

$F_1.E=[-5,-3,-5,-1,5,-1,5]^T$ which translates into $F_2 = [1, 0, 0, 0, 1, 0, 1]^T$,
$F_2.E=[-4,-4,-4,-4,4,0,8]^T$ which translates into $F_3 = [1, 0, 0, 0, 1, 0, 1]^T$.

Notice that $F_3 = F_2$, and the network converges to the limit cycle: $F_2 \rightarrow F_3 = F_2$.

Unsuccessful users who are search oriented by searching can fail again or be successful, which is a contradiction.

browse oriented and Unsuccessful Users in CE as in Input Vector

Looking at such an input vector: $B_1 = [0, 1, 0, 0, 0, 0, 1]^T$, then applying 6.13 to B_1:

$B_1.E =[-3,-1,-3,1,7,-3,3]^T$ which translates into $B_2 = [0, 1, 0, 1, 1, 0, 1]^T$
$B_2.E =[1,-1,-3,1,11,-7,7]^T$ which translates into $B_3 = [1, 1, 0, 1, 1, 0, 1]^T$
$B_3.E =[0,0,-4,4,12,-8,8]^T$ which translates into $B_4 = [0, 1, 0, 1, 1, 0, 1]^T$

Notice that $B_4 = B_2$, and the network converges to the limit cycle: $B_2 \rightarrow B_3 \rightarrow B_4 = B_2$. Unsuccessful users who are browse oriented in the CE model by browsing constrained information can fail again or be successful, which is a contradiction.

search oriented and successful Users in CE as in Input Vector

Looking at such an input vector: $S_1 = [1, 0, 0, 0, 1, 0, 0]^T$, then applying 6.13 to S_1:

$S_1.E =[-1,1,-1,3,5,-1,5]^T$ which translates into $S_2 =[1,1,0,1,1,0,1]^T$,
$S_2.E =[-1,1,-1,3,5,-1,5]^T$ which translates into $S_3 =[1,1,0,1,1,0,1]^T$.

Notice that $S_3 = S_2$, and the network converges to the limit cycle: $S_2 \rightarrow S_3 = S_2$. Successful users who are search oriented in the CE model by searching or browsing constrained information can fail or remain successful, which is a contradiction.

browse oriented and successful Users in CE as in Input Vector

Looking at such an input vector: $S_1 = [0, 1, 0, 0, 1, 0, 0]^T$, then applying 6.13 to S_1:

$S_1.E =[1,3,0,5,7,-3,3]^T$ which translates into $S_2 = [1, 1, 0, 1, 1, 0, 1]^T$,
$S_2.E =[0,0,-4,4,12,-8,8]^T$ which translates into $S_3 = [0, 1, 0, 1, 1, 0, 1]^T$,
$S_3.E =[1,-1,-3,1,11,-7,7]^T$ which translates into $S_4 = [1, 1, 0, 1, 1, 0, 1]^T$.

Notice that $S_3 = S_2$, and the network converges to the limit cycle: $S_2 \rightarrow S_3 \rightarrow S_4 = S_2$.

Successful users who are browse oriented in the CE model by searching and browsing constrained information can fail or remain successful, which is a contradiction. correlation encoding makes wrong recommendations on success and failure. Correlation encoding treats negative and zero causal edges the same. It can encode spurious causal implications between concepts such as the last rule just discovered. Correlation encoding was proven to be a poor model of inferred causality in other domains. (page 520 of [69]).

Adaptive Fuzzy Cognitive Maps of User's Behavior using differential hebbian learning

Differential Hebbian Learning (DHL) ([67]) encodes causal changes to avoid spurious causality that is seen in correlation encoding. The concepts must move in the same or opposite directions to infer a causal link. Just being on does not lead to a new causal link. DHL correlates the changes or signal velocities $C_i(x_i)'C_j(x_j)'$ and not the signals $C_i(x_i)C_j(x_j)$. The patterns of turning on or off must correlate positively or negatively:

$$e'_{i,j} = -e_{i,j} + C_i(x_i)C_j(x_j) \qquad (6.17)$$

$C_i(x_i)C_j(x_j) > 0$ iff concepts C_i and C_j move in the same direction.

$C_i(x_i)C_j(x_j) < 0$ iff C_i and C_j move in the opposite direction. Thus, e_{ij} is an exponential weighted average of lagged changes. The most recent changes have the most weight. The discrete change $\Delta C_i(t) = C_i(t) - C_i(t-1)$ lies in -1, 0, 1. The discrete differential hebbian learning can take the form:

$$e_{i,j}(t+1) = e_{i,j}(t) + t[\Delta C_i(x_i)\Delta C_j(x_j) - e_{i,j}(t)] \text{ if} \Delta C_j(t) \neq 0 \qquad (6.18)$$
$$= e_{ij}(t) \text{if} C_j(t) = 0$$

μ_t is a learning coefficient that decreases with time to help forget old causal strengths in favor of new ones:

$$\mu_t = 0.1[1 - \frac{t}{(1.1N)}] = 0.1[1 - \frac{t}{7.7}] \qquad (6.19)$$

Here is an example on how to calculate each term in the new E matrix:

$$e_{11}(t+1) = e_{11}(t) + \mu_t(\Delta[C_1(x_1)]\Delta[C_1(x_1)] - e_{11}(t))$$
$$= 0 + \mu_t(\Delta C_1(x_1)\Delta C_1(x_1) - 0)$$
$$= \mu_t(\Delta C_1(x_1)\Delta C_1(x_1)).1$$
$$= \mu_t$$

then E becomes:

$$E = \begin{array}{c} \\ C_1 \\ C_2 \\ C_3 \\ C_4 \\ C_5 \\ C_6 \\ C_7 \end{array} \begin{array}{ccccccc} C_1 & C_2 & C_3 & C_4 & C_5 & C_6 & C_7 \\ \left[\begin{array}{ccccccc} .25 & -.75 & -.75 & .5 & .75 & .75 & .75 \\ .75 & 0 & -.75 & -.75 & .75 & -.75 & .75 \\ -.75 & -.75 & 0 & .75 & 0 & .75 & .75 \\ -.5 & -.5 & -.5 & .25 & .25 & .25 & .25 \\ .75 & .75 & .5 & .75 & .25 & .5 & -.75 \\ .75 & .75 & -.25 & .75 & .75 & -.25 & -.75 \\ .5 & -.5 & -.5 & .75 & .25 & -.25 & .25 \end{array}\right] \end{array}$$

search oriented and Unsuccessful Users in DHL as in Input Vector
Looking at such an input vector: $F_1=[1,0,0,0,0,0,1]^T$ with a threshold equal
to 0.49, and relaxing the condition of searching afterwards:

$F_1.E=[0,-2,-1.25,2,1,1.25,1.75]^T$ which translates to $F_2=[0,0,0,1,1,1,1]^T$
$F_2.E=[1.5,.5,-.75,2.5,1.5,.25,-1]^T$ which translates to $F_3=[1,1,0,1,1,0,0]^T$
$F_3.E=[1.25,-.5,-1.5,.75,2,.75,1]^T$ which translates to $F_4=[1,0,0,1,1,1,1]^T$
$F_4.E=[1.75,-.25,-1.5,3,2.25,1,-.25]^T$ which translates to $F_5=[1,0,0,1,1,1,0]^T$
$F_5.E=[1.25,.25,-1,2.25,2,1.25,-.5]^T$ which translates to $F_6=[1,0,0,1,1,1,0]^T$

Notice that $F_6 = F_5$ and the network converges to the limit cycle $F_5 \rightarrow F_6$.
Unsuccessful users who are search oriented by searching constrained (enough
experience) but relevant information can turn their failure into success.
browse oriented and Unsuccessful Users in DHL as in Input Vector
Looking at such an input vector: $B_1=[0,1,0,0,0,0,1]^T$ with a threshold equal
to 0.49, and relaxing the condition of browsing afterwards, then $B_1.E =[.5,-1.25,-1.25,.75,1,-.25,1.75]$, which translates into $B_2=[1\ 0\ 0\ 1\ 1\ 0\ 1]$,

$B_2.E=[1,-1,-1.25,2.25,1.5,1.25,.5]$, which translates into $B_3=[1,0,0,1,1,1,1]$
$B_3.E=[1.75,-.25,-1.5,3,2.25,1,-.25]$, which translates into $B_4=[1,0,0,1,1,1,0]$
$B_4.E=[1.25,.25,-1,2.25,2,1.25,-.5]$, which translates into $B_5=[1,0,0,1,1,1,0]$

Notice that $B_5 = B_4$ and the network converges to the limit cycle $B_4 \rightarrow B_5$.
Unsuccessful users who are browse oriented by searching constrained (enough
experience) but relevant information can turn their failure into success.
search oriented and successful Users in DHL as in Input Vector
Looking at such an input vector: $S_1=[1,0,0,0,1,0,0]$ with a threshold equal to
0.49 and relaxing the search constraints afterwards, then:

$S_1.E=[.25,-.75,-.25,2.0,1.0,2.0,.75]$, which translates into $S_2=[0,0,0,1,1,1,1]$
$S_2.E=[1.5,.5,-.75,2.5,1.5,.25,-1.0]$, which translates into $S_3=[1,1,0,1,1,0,0]$
$S_3.E=[2.0,.25,-1.75,1.5,2.75,0.5,.25]$, which translates into $S_4=[1,0,0,1,1,1,0]$
$S_4.E=[1.25,.25,-1.0,2.25,2.0,1.25,-.5]$, which translates into $S_5=[1,0,0,1,1,1,0]$

Notice that $S_5=S_4$ and the network converges to the limit cycle $S_4 \rightarrow S_5=S_4$.
Successful users who are search oriented in the DHL model by searching constrained but relevant information can remain successful.
browse oriented and successful Users in DHL as in Input Vector
Looking at the above as an input vector: $S_1=[0,1,0,0,1,0,0]$, with a threshold
equal to 0.49 and relaxing the search constraints afterwards, then:

$S_1.E=[.75,0,-.25,.75,1.0,.5,.75]$ which translates into $S_2 =[1,0,0,1,1,1,1]$,

$S_2.E=[1.75,-.25,-1.5,3.0,2.25,1.0,-.25]$ which translates into $S_3 =[1,0,0,1,1,1,0]$,
$S_3.E=[1.25,.25,-1.0,2.25,2.0,1.25,-.5]$ which translates into $S_4 =[1,0,0,1,1,1,0]$.

Notice that $S_4=S_3$ and the network converges to the limit cycle $S_3 \to S_4 = S_3$.
Successful users who are browse oriented in the DHL model by searching constrained but relevant information can remain successful.

The Case of Negative Influence of C_7 on the Other Concepts

$$E = \begin{array}{c} \\ C_1 \\ C_2 \\ C_3 \\ C_4 \\ C_5 \\ C_6 \\ C_7 \end{array} \begin{array}{c} \begin{array}{ccccccc} C_1 & C_2 & C_3 & C_4 & C_5 & C_6 & C_7 \end{array} \\ \left[\begin{array}{ccccccc} 0 & -1 & -1 & 1 & 1 & -1 & 1 \\ 1 & 0 & -1 & -1 & 1 & -1 & 1 \\ -1 & -1 & 0 & -1 & 0 & 1 & 1 \\ -1 & -1 & -1 & 0 & 0 & 0 & 0 \\ 1 & 1 & 1 & 1 & 0 & 1 & -1 \\ 1 & 1 & 0 & 1 & 1 & 0 & -1 \\ -1 & -1 & -1 & -1 & 0 & 0 & 0 \end{array} \right] \end{array}$$

Figure 6.10 visualizes the influences among the concepts. (only negative influences are shown). For example, it says that C_7 has no influence (no arrow) over C_7, has negative influence over C_1, C_2, C_3, and C_4.

search oriented and Unsuccessful Users in FCM as in Input Vector
Looking at such an input vector: $F_1=[1,0,0,0,0,0,1]$, and applying 6.13 to F_1:

$F_1.E =[-2,-3,-2,-1,1,0,2]$, which translates into $F_2=[1,0,0,0,1,0,1]$,
$F_2.E =[-1,-2,-1,0,1,1,1]$, which translates into $F_3 =[1,0,0,0,1,1,1]$,
$F_3.E =[0,-1,-1,1,2,1,0]$, which translates into: $F_4 =[1,0,0,1,1,1,0]$,
$F_4.E =[0,-1,-1,2,2,1,0]$, which translates into $F_5 =[1,0,0,1,1,1,0]$.

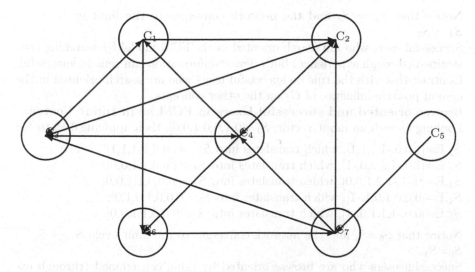

Fig. 6.10. FCM of User's experience in the case of negative influence of C_7

Notice that $F_5 = F_4$, and the network converges to the limit cycle: $F_4 \rightarrow F_5 = F_4$ Unsuccessful users who are search oriented in the FCM model by searching constrained (through experience) but relevant information can turn their failure into success. Contrast this with the rule on unsuccessful users who are browse oriented in the case of positive influence of C_7 on the other concepts.

browse oriented and Unsuccessful Users in FCM as in Input Vector

Looking at such an input vector: $F_1=[0,1,0,0,0,0,1]$, and applying 6.13 to F_1:

$F_1.E = [-1,-2,-2,-3,1,0,2]$, which translates into $F_2=[0,1,0,1,1,0,1]$,
$F_2.E = [1,-1,-1,-2,1,0,0]$, which translates into $F_3=[1,1,0,0,1,0,0]$,
$F_3.E = [1,0,-1,-1,2,1,0]$, which translates into $F_4=[1,1,0,1,1,0,1]$,
$F_4.E = [0,0,-1,-1,3,0,0]$, which translates into $F_5=[0,1,0,0,1,0,1]$,
$F_5.E = [1,-3,-3,-1,2,0,2]$, which translates into $F_6=[1,1,0,0,1,0,0]$.

Notice that $F_2 = F_6$, and the network converges to the limit cycle: $F_3 \rightarrow F_4 \rightarrow F_5 \rightarrow F_6=F_3$.

Unsuccessful users who are browse oriented in the FCM model by searching or browsing can turn their failure into success. Contrast this with the rule on unsuccessful users who are browse oriented in the case of positive influence of C_7 on the other concepts.

search oriented and successful Users in FCM as in Input Vector

Looking at such an input vector: $S_1=[1,0,0,0,1,0,0]$, then applying 6.13 to S_1:

$S_1.E=[0,-1,0,1,1,1,1]$, which translates into $S_2 =[1,0,0,1,1,1,1,1]$,
$S_2.E=[-1,-2,-2,1,2,1,0]$, which translates into $S_3=[1,0,0,1,1,1,0]$,
$S_3.E=[-1,-2,-2,1,2,1,0]$, which translates into $S_4=[1,0,0,1,1,1,0]$,
$S_4.E=[-1,-2,-2,1,2,1,0]$, which translates into $S_5=[1,0,0,1,1,1,0]$,
$S_5.E=[0,-1,-1,2,2,1,0]$, which translates into $S_6=[1,0,0,1,1,1,0]$.

Notice that $S_4 = S_3$ and the network converges to the limit cycle $S_3 \rightarrow S_4 = S_3$

Successful users who are search oriented in the FCM model by searching constrained (through experience) but relevant information can remain successful. Contrast this with the rule on successful users who are search oriented in the case of positive influence of C_7 on the other concepts.

browse oriented and successful Users in FCM as in Input Vector

Looking at such an input vector: $S_1=[0,1,0,0,1,0,0]$, then applying 6.13 to S_1:

$S_1.E=[1,0,0,-1,1,1,1]$, which translates into $S_2 =[1,0,0,0,1,1,1]$,
$S_2.E=[1,0,-1,2,2,0,-1]$, which translates into $S_3=[1,0,0,1,1,0,0]$,
$S_3.E=[0,-1,-1,2,1,0,0]$, which translates into $S_4 =[0,0,0,1,1,0,0]$,
$S_4.E=[0,0,0,1,0,1,-1]$, which translates into $S_5=[0,0,0,1,0,1,0]$,
$S_5.E=[0,0,-1,1,1,0,-1]$, which translates into $S_6=[0,0,0,1,1,0,0]$.

Notice that $S_6=S_4$ and the network converges to the limit cycle $S_4 \rightarrow S_5 \rightarrow S_6=S_4$.

Successful users who are browse oriented by using constrained (through experience) but relevant information (S5) can remain successful. The network has a limit cycle of length 2. The limit cycle should be less than the number of concepts otherwise crosstalk can occur. Contrast this with the rule on successful users who are browse oriented in the case of positive influence of C_7 on the other concepts.

Adaptive Fuzzy Cognitive Maps through Correlation of User's Web Behavior

Would correlation encoding applied earlier behave any better in the case of negative influence of C_7 on the other concepts or should we expect the same kind of behavior experienced above? To answer such a question an adaptive FCM is built that uses correlation to encode some limit cycles in the FCMs or temporal associative memories (TAM). Correlation encoding method can store only few patterns. To encode binary limit cycles in matrix E the TAM method sums the weighted correlation matrices between successive states. Applying (6.16) to the new matrix E yields:

$$
E = \begin{array}{c} \\ C_1 \\ C_2 \\ C_3 \\ C_4 \\ C_5 \\ C_6 \\ C_7 \end{array}
\begin{array}{c} C_1 \ \ C_2 \ \ C_3 \ C_4 \ C_5 \ C_6 \ C_7 \end{array}
\left[\begin{array}{ccccccc}
-1 & 1 & -1 & -1 & -1 & 1 & -1 \\
1 & 3 & 1 & 1 & 1 & -1 & -3 \\
3 & 5 & 3 & 3 & 3 & 1 & -1 \\
3 & 1 & -1 & -1 & 3 & -3 & -1 \\
-1 & -3 & -1 & -5 & -1 & 1 & 3 \\
1 & 3 & 1 & 1 & 1 & -1 & -3 \\
-1 & -3 & -1 & -5 & -1 & 1 & 3
\end{array} \right]
$$

Looking at such an input vector: $F_1=[1,0,0,0,0,0,1]$, then applying (6.13) to F_1 :

$F_1.E=[1,3,0,-3,1,3,1]$, which translates into $F_2=[1,1,0,0,1,1,1]$
$F_2.E=[2,6,0,-6,2,2,-2]$, which translates into $F_3=[1,1,0,0,1,1,0]$
$F_3.E=[3,9,0,-1,3,1,-5]$, which translates into $F_4=[1,1,0,0,1,1,0]$

Notice how $F_4 = F_3$, and the network converges to the limit cycle $F_3 \rightarrow F_4=F_3$.

Unsuccessful users who are search oriented in the CE model by searching or browsing relevant information can turn their failure into success. Contrast this with the rule on unsuccessful users who are search oriented in the case of positive influence of C_7 on the other concepts.

browse oriented and Unsuccessful Users in CE as in Input Vector

Looking at such an input vector: $B_1 = [0,1,0,0,0,0,1]$, then by applying (6.13) to B_1:

$B_1.E=[3,5,0,-1,3,1,-1]$, which translates into $B_2=[1,1,0,0,1,1,0]$,
$B_1.E=[3,9,0,-1,3,1,-5]$, which translates into $B_3=[1,1,0,0,1,1,0]$.

Notice how $B_3 = B_2$, and the network converges to the limit cycle $B_2 \rightarrow B_3=B_2$.

Unsuccessful users who are browse oriented in the CE model by searching or browsing relevant information can turn their failure into success. Contrast this with the rule on unsuccessful users who are browse oriented in the case of positive influence of C_7 on the other concepts.

search oriented and successful Users in CE as in Input Vector
Looking at such an input vector: $S_1=[1,0,0,0,1,0,0]$, then applying (6.13) to S_1:

$S_1.E =[1,3,-1,-3,1,3,1]$, which translates into $S_2=[1,1,0,0,1,1,1]$,
$S_2.E =[2,6,-2,-6,2,2,-2]$, which translates into $S_3=[1,1,0,0,1,1,0]$,
$S_3.E =[3,9,-3,-1,3,1,-5]$, which translates into $S_4=[1,1,0,0,1,1,0]$.

Notice how $S_4 = S_3$, and the network converges to the limit cycle $S_3 \rightarrow S_4 = S_3$ Successful users who are search oriented in the CE model by searching or browsing relevant information remain successful. Contrast this with the rule on successful users who are search oriented in the case of positive influence of C_7 on the other concepts.

browse oriented and successful Users in CE as in Input Vector
Looking at such an input vector: $B_1=[0,1,0,0,1,0,0]$, then applying (6.13) to B_1

$B_1.E =[3,5,-3,-1,3,1,-1]$, which translates into $B_2=[1,1,0,0,1,1,0]$,
$B_2.E =[3,9,-3,-1,3,1,-5]$, which translates into $B_3=[1,1,0,0,1,1,0]$.

Notice how $B_3 = B_2$, and the network converges to the limit cycle $B_2 \rightarrow B_3 = B_2$

Successful users who are browse oriented in the CE model by searching or browsing relevant information remain successful. The case of negative influence of C_7 on other concepts does not seem to yield the same contradictory results that the case of positive influence of C_7 on the other concepts yielded above.

Adaptive Fuzzy Cognitive Maps of User's Behavior using differential hebbian learning

$$e_{11}(t + 1) = e_{11}(t) + \mu_t(\Delta[C_1(x_1)]\Delta[C_1(x_1)] - e_{11}(t)]$$
$$= 0 + \mu_t(\Delta C_1(x_1)\Delta C_1(x_1) - 0)$$
$$= \mu_t(\Delta C_1(x_1)\Delta C_1(x_1)).1$$
$$= \mu_t$$

then E becomes:

$$
E = \begin{array}{c|ccccccc}
 & C_1 & C_2 & C_3 & C_4 & C_5 & C_6 & C_7 \\
\hline
C_1 & .25 & -.75 & -.75 & .5 & .75 & .75 & .75 \\
C_2 & .75 & 0 & -.75 & -.75 & .75 & -.75 & .75 \\
C_3 & -.75 & -.75 & 0 & .75 & 0 & .75 & .75 \\
C_4 & -.5 & -.5 & -.5 & .25 & .25 & .25 & .25 \\
C_5 & .75 & .75 & .5 & .75 & .25 & .5 & -.75 \\
C_6 & .75 & .75 & 0 & .75 & .75 & 0 & -.75 \\
C_7 & -.75 & -.75 & -.75 & -.5 & 0 & 0 & .25
\end{array}
$$

search oriented and Unsuccessful Users in DHL as in Input Vector
Looking at the above as an input vector: $F_1=[1,0,0,0,0,0,1]$ with a threshold equal to 0.49, and by relaxing the searching constraint condition afterwards:

$F_1.E=[-1.25,-2.25,-1.5,.75,.75,1.5,1.75]$ which translates into $F_2=[0,0,0,1,1,1,1]$
$F_2.E=[.25,.25,-.75,1.25,1.25,.75,-1.0]$ which translates into $F_3=[0,0,0,1,1,1,0]$
$F_3.E=[1.0,1.0,0,0,1.75,1.25,.75,-1.25]$ which translates into $F_4=[1,1,0,1,1,1,0]$
$F_4.E=[2.0,.25,-1.5,1.5,2.75,.75,.25]$ which translates into $F_5=[1,0,0,1,1,1,0]$
$F_5.E=[1.25,.25,-.75,2.25,2.0,1.5,-.5]$ which translates into $F_6=[1,0,0,1,1,1,0]$.

Notice that $F_6=F_5$ and the network converges to the limit cycle $F_5 \rightarrow F_6=F_5$. Unsuccessful users who are search oriented in the DHL model by searching constrained (through experience) but relevant information can turn their failure into success. Contrast this with the rule on unsuccessful users who are search oriented in the case of positive influence of C_7 on the other concepts.
browse oriented and Unsuccessful Users in DHL as in Input Vector
Looking at the above as an input vector: $F_1=[0,1,0,0,0,0,1]$ with a threshold equal to 0.49, and by relaxing the browsing condition afterwards, then:

$F_1.E =[-.75,-1.5,-1.5,-.5,.75,0,1.75]$, which translates into $F_2 =[0,0,0,0,1,0,1]$,
$F_2.E =[0,0,-.25,.25,.25,0.5,-.5]$, which translates into $F_3=[0,0,0,0,0,1,0]$,
$F_3.E =[0.75,.75,0,.75,.75,0,-.75]$, which translates into $F_4=[1,1,0,1,1,0,0]$,
$F_4.E =[1.25,-0.5,-1.5,.75,2.0,.75,1.0]$, which translates into $F_5=[1,0,0,1,1,1,1]$,
$F_5.E=[.5,-.5,-1.5,1.75,2.0,1.5,-.25]$, which translates into $F_6=[1,0,0,1,1,1,0]$,
$F_6.E =[1,0,0,1,1,1,0]$ which translates into $F_7=[1,0,0,1,1,1,0]$.

Notice that $F_6=F_7$ and the network converges to the limit cycle $F_6 \rightarrow F_7=F_6$. Unsuccessful users who are browse oriented in the DHL model by searching constrained (through experience) but relevant information can turn their failure into success. Contrast this with the rule on unsuccessful users who are browse oriented in the case of positive influence of C_7 on the other concepts.
search oriented and successful Users in DHL as in Input Vector
Looking at the above as an input vector: $S_1=[1,0,0,0,1,0,0]$, and by relaxing the searching constraint condition afterwards, then:

$S_1.E =[.25,-.75,-.25,2.0,1.0,2.0,.75]$, which translates into $S_2=[0,0,0,1,1,1,1]$,
$S_2.E =[.25,.25,-.75,1.25,1.25,.75,-1.0]$, which translates into $S_3=[0,0,0,1,1,1,0]$,
$S_3.E =[1.0,1.0,0,0,1.75,1.25,.75,-1.25]$, which translates into $S_4=[1,1,0,1,1,1,0]$,
$S_4.E =[2.0,.25,-1.5,1.5,2.75,.75,.25]$, which translates into $S_5=[1,0,0,1,1,1,0]$,
$S_5.E =[1.25,.25,-.75,2.25,2.0,1.5,-.5]$, which translates into $S_6=[1,0,0,1,1,1,0]$.

Notice that $S_6=S_5$ and the network converges to the limit cycle $S_5 \rightarrow S_6=S_5$. Successful users who are search oriented in the DHL model by searching constrained (through experience) but relevant information can remain successful. Contrast this with the rule on successful users who are search oriented in the case of positive influence of C_7 on the other concepts.
browse oriented and successful Users in DHL as in Input Vector:
Looking at the above as an input vector: $S_1=[0,1,0,0,1,0,0]$, then:

$S_1.E =[.75,0,-.25,.75,1.0,.5,.75]$, which translates into $S_2=[1,0,0,1,1,1,1]$,
$S_2.E =[.5,-.5,-1.5,1.75,2.0,1.5,-.25]$, which translates into $S_3 =[1,0,0,1,1,1,0]$,
$S_3.E =[1.25,.25,-.75,2.25,2.0,1.5,-.5]$, which translates into $S_4 =[1,0,0,1,1,1,0]$.

Notice that $S_4 = S_3$ and the network converges to the limit cycle $S_3 \rightarrow S_4 = S_3$ Successful users who are browse oriented in the DHL model by searching constrained (through experience) but relevant information can remain successful. Contrast this with the rule on successful users who are browse oriented in the case of positive influence of C_7 on the other concepts.

6.3.4 Final Observation on FCM as a Model for User's Web Behavior

Fuzzy cognitive maps can model the causal graph of expert's perception of how users behave on the Web. Although experts are still divided on the reasons users fail to answer a query, we establishes a stable dynamical system that behaves well in a limit cycle. Tables 6.9 and 6.10 summarizes the rules that were uncovered in the case of success and failure in different models. When asked what happens in the case of an unsuccessful or successful user who is searching, the original FCM causal model, with the positive influence of C_7 on other concepts, responds well and moves towards an attractor with a limit cycle and gives some meaningful rules. When asked what happens in the case of an unsuccessful or successful user who is browsing, the original causal FCM model, with the positive influence of C_7 on other concepts, moves towards an attractor with a limit cycle but gives some contradictory rules. When asked what happens in the case of an unsuccessful or successful user who is searching

Table 6.9. Rules on success uncovered by the original model, CE, and DHL model(C.=Constrained, R.I.=Relevant Information)

Influence Of C7	Algorithms	SS (Successful and search oriented)	SB (Successful and browse oriented)
Positive	FCM	By searching C, but R.I. remain successful	By searching/browsing C but R.I. remain successful or fail.
	CE	By searching or browsing C.I. remain successful or fail	By searching/browsing C.I. fail or remain successful
	DHL	By searching C. but R.I. remain successful	By searching C. but R.I. remain successful
Negative	FCM	By searching C. but R.I. remain successful	By using C. but R.I. remain successful
	CE	By searching/ browsing C.I. remain successful	By searching/ browsing C.I. remain successful
	DHL	By searching C. but R.I. remain successful	By searching C. but R.I. remain successful

Table 6.10. Rules on failure uncovered by the original model, CE, and DHL model.(C.= Constrained, R.I.= Relevant Information)

Influence of $C7$	Algorithms	FS (Failed and search oriented)	FB (Failed and browse oriented)
Positive	FCM	By searching C. but R.I. turn failure into success.	By searching/browsing C. but R.I. become successful or remain unsuccessful
	CE	By searching stay unsuccessful or become successful.	By browsing C.I. remain unsuccessful or become successful.
	DHL	By searching C. but R.I.turn failure into success.	By searching C. but R.I. turn failure into success.
Negative	FCM	By searching C. but R.I. turn failure into success.	By searching/browsing C. but R.I. turn failure into success
	CE	By searching/ browsing R.I. turn failure into success.	By browsing/searching C.I. turn failure into success.
	DHL	By searching C. but R.I.turn failure into success.	By searching C. but R.I. turn failure into success.

or browsing, the original FCM causal model, with the negative influence of C_7 on other concepts, responds well and moves towards an attractor with a limit cycle and gives some meaningful rules. This signifies that the original FCM model with the negative influence of C_7 on other concepts is a better cognitive model to uncover causal rules between the different concepts. When the original FCM causal system is left to behave in time, does the system learn any new rules or reinforce old ones? To answer the question, 2 models were considered:

1. If differential hebbian learning is used (DHL), the system uncovers new rules in the case of failure and in the case of success. That shows that the original causal matrix can uncover some dynamical rules on learning when left to behave in time, which was not apparent in the beginning as follows: when asked what happens in the case of a unsuccessful or successful user who is searching or browsing, the DHL model, in the case of positive or negative influence of C_7 on other concepts, responds well and moves towards an attractor with a limit cycle and gives some meaningful rules. Not only that, but the dynamical system can move from failure to success with certain conditions imposed on the user. The same rules that switch users from failure to success can keep successful users successful. The same rule discovered in the case of positive influence of C_7 is rediscovered in the case of negative influence of C_7 in the DHL model, which shows that the rule becomes an invariant in the system regardless of the positive or negative influences of C_7.

2. If correlation encoding (CE) is used, the system behaves erratically in the case of positive influence of C_7 on other concepts and produces contradictory rules. In the case of negative influence of C_7 on other concepts the dynamical system behaves as follows: when asked what happens in the case of a unsuccessful or successful user who is searching or browsing, the CE model, in the case of negative influence of C_7 on other concepts, responds well and moves towards an attractor with a limit cycle and gives some meaningful rules. Not only that, but the dynamical system can move from failure to success with certain conditions imposed on the user. The same rule that switch users from failure to success can keep successful users successful. The CE model did not produce any invariant rules. That confirms the fact that the original causal model with negative influence of C_7 on other concepts is a better cognitive model with time.

What if the concepts of searching or browsing were allowed different values in them, since the results show that users did not search or browse equally: some searched or browsed once, some searched or browsed an average number of times, and some used a great number of searches or browses before they exited the system. If searching was divided into 3 different categories: Little searching (LS), average searching (AS) and heavy searching (HS), Figure 6.11 shows the concept search broken into 3 different categories. Nested FCM can be used in that case where the final matrix E is the sum of all the matrices for all the 3 categories:

$$E = \frac{(E_{LS} + E_{AS} + E_{HS})}{3} \tag{6.20}$$

The same thing can be applied to browsing, since not all users browse equally. Figure 6.12 shows the concept "browse" divided into 3 different

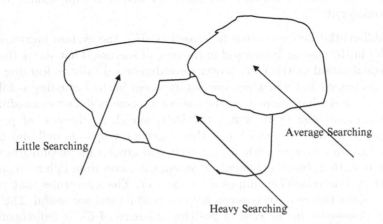

Fig. 6.11. Nested FCMs can divide the concept Search into 3 categories

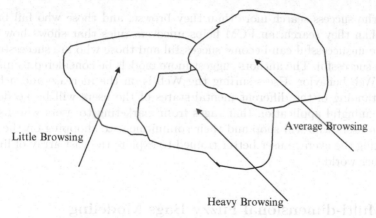

Fig. 6.12. Nested FCMs can divide the concept Browsing into 3 categories

categories: Little browsing (LB), average browsing (AB), and heavy browsing (HB). How does the original FCM behave in the case of the 3 categories

$$E = \frac{(E_{LB} + E_{AB} + E_{HB})}{3} \qquad (6.21)$$

Would nested FCMs help improve the overall behavior of the system and avoids the chaotic behavior of the system? The answer to such a question lies in the fact that more studies reflecting the fine grain division of the concept of searching or browsing into three 3 categories and how the overall behavior of successful users and unsuccessful users can help design these nested FCM. Remember that an FCM is the expert's-centered view on how users search not a user-centered view on how they do search or browse.

relevance. If a user loops while searching or loops while browsing, he/she stresses out because of Time constraints and information overload, thus he/she fails. In the case of success, if a user loops while searching or loops while browsing, stresses out because of Time constraints and information overload but uncovers relevant information, he/she succeeds. This model shows that relevance is a key issue in separating success from failure. The same information presented to a user can be interpreted as being relevant to a user while it is being interpreted irrelevant to another. relevance becomes a major issue in information interpretation and filtering. relevance needs to be exploited as part of a query expansion and relevance feedback to improve information retrieval (*Chang & Hsu* [21]). Popularity should not be confused with relevance. Although google and other search engines use different ways to judge relevance and ranking (*Meghabghab* [85]), this is still a very relevant issue and more research in this regard to provide a reliable automatic relevance algorithm will help improve returned set results and reduce the high failing rate of users (40% in our case) in searching for information. When compared to the markovian modeling, the dynamics of the FCM model reveals things never suspected before. While the markovian model reveals general characteristics that users who success search more than they browse, and those who fail browse more than they search, an FCM helps uncovers rules that shows how users who are unsuccessful can become successful and those who are successful can remain successful. The authors suggest more models be considered to uncover user's Web behavior. Users surfing the Web is on the increase and a better understanding of the different mental states of the users will be needed for any meaningful application that rages from marketing to users who use the Web, to users who will shop and their commitment to shopping on the Web, to making the average user better trained to explore the vast array of links in the cyber world.

6.4 Multi-dimensional Fuzzy Bags Modeling User's Web Behavior

In all studies of user's behavior on searching the Web, researchers found that their search strategy or traversal process showed a pattern of moving back and forth between searching and browsing. These activities of searching and browsing are inclusive activities to complete a goal-oriented task to find the answer for the query. The moving back and forth, including backtracking to a former Web page, constitutes the measurable or physical states or actions that a user takes while searching the Web. How do we account for the repeated searching and browsing and shifting between different Web pages to better understand the behavior of users? Is there a mathematical model that can help model actions of users to distinguish between 2 users while their repeated actions are accounted for since it is a part of their behavior or inherent essence to arrive at the goal of answering the questions regardless whether they are successful or unsuccessful? The idea of a set does not help account for repeated actions since in a set, repeated objects are ignored. In some situations we want a structure in which a collections of objects in the same sense as a set but a redundancy counts. The structure of a bag as a framework can help study the behavior of users and uncover some intrinsic properties about users. Also would bags help distinguish between 2 sets of users based on their syntactic properties without a priori knowledge of their semantic properties? This study constitutes a premier in applying bags to user's mining of the Web. The research will consider queries that depend on one variable or "uni-variable". An example of "uni-variable" queries is: "Find the number of Web pages of users whose searching is less than the average". Crisp bags failed to answer such a query. Fuzzy operators were used successfully in comparing the "fuzzy bag" of the successful users to the "fuzzy bag" of unsuccessful users to answer such a query. A distance between 2 fuzzy bags was also established. The research also considered queries that depend on more than one variable or multi-variable. An example of a "multi-variable" query is: "Find the number of Web pages of users whose number of searches is less than average and hyperlink navigation is less than the average". One-dimensional fuzzy bags were extended to two dimensional fuzzy bags and two dimensional fuzzy bag operators were used to answer such a query.

6.4.1 Introduction

In all studies of user's behavior on searching the Web, researchers found that their search strategy or traversal process showed a pattern of moving back and forth between searching and browsing. These activities of searching and browsing are inclusive activities to complete a goal-oriented task to find the answer for the query [62]. The moving back and forth, including backtracking to a former Web page, constitutes the measurable or physical states or actions that a user takes while searching the Web [36]. How do we account for the repeated searching and browsing and shifting between different Web pages to better understand the behavior of users? Is there a mathematical model that can help model actions of users to distinguish between 2 users while their repeated actions are accounted for since it is a part of their behavior or inherent essence to arrive at the goal of answering the questions regardless whether they are successful or unsuccessful? DiLascio et al. [28] described the mental state of the user in an Intelligent Tutoring System (ITS) while the user is navigating the system to arrive at an answer. The navigation module is a key since it is responsible for managing the effects of disorientation and cognitive overhead. The mental state of the user is described by a set of triples $< CS, PS, OR >$ where:

Table 6.11. The greatest fuzzy bag such that Max of $w_{(S_1)_\alpha}(1W) + w_{(A_S \cap A_F)_\alpha}(1W) \leq w_{(AS)_\alpha}(1W)$

α	$w_{A_S\alpha}(1w)$	$w_{A_F\alpha}(1w)$	$w_{A_S\alpha}(1w)$ $-w_{A_F\alpha}(1w)$	Max of $w_{S_{1\alpha}}(1w)$	$w_{(A_S\cap A_F)\alpha}(x)$
$< .9 >$	3	1	2	1	1
$< .8 >$	4	2	2	1	2
$< .7 >$	4	3	1	2	3
$< .6 >$	5	3	2	2	3

Table 6.12. The greatest fuzzy bag such that Max of $w_{(S_1)_\alpha}(1W) + w_{(A_S\cap A_F)_\alpha}(1W) \leq w_{(AS)_\alpha}(1W)$

α	$w_{A_S\alpha}(1w)$	$w_{A_F\alpha}(1w)$	$w_{A_S\alpha}(1w)$ $-w_{A_F\alpha}(1w)$	Max of $w_{S_{1\alpha}}(1w)$	$w_{(A_S\cap A_F)\alpha}$
$< 1,1 >$	3	1	2	2	1
$< .9,.8 >$	3	2	1	1	2
$< .9,.6 >$	4	2	2	2	2
$< .8,.9 >$	4	3	1	1	3
$< .7,.4 >$	5	3	2	2	3

Table 6.13. $S_1(1W) = \{ \ll 1,1 >, < .7,.4 \gg //1W \}$ is the largest fuzzy bag where $w_{(S_1)}\alpha(1W) \leq w_{(S_1)\alpha}(1W)$

α	$w_{A_S\alpha}(1w)$	$w_{A_F\alpha}(1w)$	$w_{A_S\alpha}(1w)$ $-w_{A_F\alpha}(1w)$	Max of $w_{S_{1\alpha}(1w)}$	$w_{(A_S \cap A_F)\alpha}$
$< 1,1 >$	3	1	2	1	2
$< .9,.8 >$	3	2	1	1	1
$< .9,.6 >$	4	2	2	1	2
$< .8,.9 >$	4	3	1	1	1
$< .7,.4 >$	5	3	2	2	2

- CS stands for cognitive states. It accounts for shifting back and forth, for comparing at any time the distance between the actual state of the user and the goal in answering the query, and for assessing the cognitive change in doing so.
- PS stands for current psychological state. It accounts for attention span and interest of the user in answering the query. These are keys in keeping the user motivated to answer the query.
- OR stands for the orientation (OR). Since a user can get frustrated because of lack of orientation and cognitive overhead. OR should be adequately measured and managed.

DiLascio et al. ([28]) did not however compare the fuzzy management of user navigation in the case of success and the case of failure of users to a given query. This research will consider such an idea. The idea of a set does not help account for repeated actions. It is known that in a set, repeated objects are ignored. In some situations we want a structure in which a collections of objects in the same sense as a set but a redundancy counts. Assume X is a set of elements that represent the actions of users while searching the Web:

- users search or query a search engine using a single concept or a multiple concept query. Let S represent such an action,
- users browse one or more categories: let B be such an action,
- users scroll results and navigate hyperlinks: let H represent such an action,
- users backtrack between different Web pages (looping can be considered a backtracking over the same Web page): let BT represent such an action,
- users select a Web page looking for an answer for the query: let W represent such an action.

Thus X as a set can be represented by X={S,B,H,BT,W} .
In section 6.4.2, we use "crisp bags" to model user's actions. In section 6.4.3, we use "fuzzy bags" to model user's actions and answer the "uni-variable" query. In section 6.4.4, we use "Multi-dimensional fuzzy bags" to model user's actions and answer the "multi-variable" query. In section 6.4.5, we summarize our findings and compare all the models used, i.e., crips bags, fuzzy bags, and difference of two Multidimensional fuzzy bags.

6.4.2 "Crisp Bag" Modeling User's Actions([86])

Def: A crisp bag A by definition [131, 132] is a count function which counts how many times a given action from $X = \{x_1, x_2, x_3, \cdots, x_n\}$ has been performed:

$$Count_A : X \to N \tag{6.22}$$

Such a definition yields to the bag A being represented by:

$A = \; < a_1/x_1, a_2/x_2, a_3/x_3, \cdots, a_n/x_n >$ where $a_i = Count_A(x_i)$.

Example. Let us look at the account of actions taken by an unsuccessful user labeled $user_{20}$ (See table 6.14 through table 6.17). User$_{20}$ queries the Web with "Limbic functions of the brain", browses the results, does not seem satisfied, queries the Web with the same query: "limbic functions of the brain", browses the results again, changes the query to "limbic system", chooses a Web page

Table 6.14. Detailed Actions of Successful Users

User	Set of Actions
User2	$< S1, B, W3, H, B >$
User2	$< 1/S, 2/B, 1/H, 0/BT, 1/W >$
User3	$< S3, H, S4, H, S5, H, BT, S5, W4, H, S5, W4, H, S6, H,$ $W3, S7, W5 >$
User3	$< 7/S, 0/B, 6/H, 1/BT, 4/W >$
User6	$< S5, H, 15, W11, S5, W11, S15, H, S5, H, W12, BT, W12 >$
User6	$< 5/S, 0/B, 3/H, 1/BT, 4/W >$
User10	$< S6, H, W8, S6, W1, BT, S6, W8 >$
User10	$< 3/S, 0/B, 1/H, 1/BT, 3/W >$
User11	$< S18, H, S19, H, BT, H, S18, H, S6, W8 >$
User11	$< 4/S, 0/B, 4/H, 1/BT, 1/W >$
User12	$< S19, W15, S19, W8, BT, S19, H, W2 >$
User12	$< 3/S, 0/B, 1/H, 1/BT, 3/W >$
User13	$< S6, W14, B, BT, S6, W1, BT, B, S6, B, W3, S6, B, S20, H,$ $S6, W3 >$
User13	$< 6/S, 4/B, 1/H, 2/BT, 4/W >$
User14	$< S17, H, W13, S6, W8, BT, S6, W14, BT, W16, BT, S6, W1,$ $S6, W17, S6, W3, S17, H, S6, W18, BT, H, W19 >$
User14	$< 8/S, 0/B, 3/H, 4/BT, 9/W >$
User15	$< S15, B, H, W19 >$
User15	$< 1/S, 1/B, 1/H, 0/BT, 1/W >$
User16	$< S3, H, B, H, BT, S6, W8, H >$
User16	$< 2/S, 1/B, 3/H, 1/BT, 1/W >$
User17	$< S1, W3, H >$
User17	$< 1/S, 0/B, 1/H, 0/BT, 1/W >$
User18	$< S6, W8, H, S6, W8, H >$
User18	$< 2/S, 0/B, 2/H, 0/BT, 2/W >$

Table 6.15. Detailed Actions of UnSuccessful Users

User	Set of Actions
User1	$< S1, H, W2, S1, H, W2, S2, BT, W1, H, S2, W2, BT, W1, BT,$ $W1, BT, W2 >$
User1	$< 4/S, 0/B, 3/H, 4/BT, 7/W >$
User4	$< S7, H, W6, S8, H, W6, S9, H, S8, H, S7, W6, S9, H, S9, W7,$ $S10, H, S11, H, S12, W8 >$
User4	$< 10/S, 0/B, 7/H, 0/BT, 5/W >$
User5	$< S3, H, B, H, S13, H, W9, BT, W10, H, S14, W2, BT, W10, H >$
User5	$< 3/S, 1/B, 5/H, 2/BT, 4/W >$
User7	$< S16, H, B, H, S17, H, W13 >$
User7	$< 2/S, 1/B, 3/H, 0/BT, 1/W >$
User8	$< S15, H, W2, S15, H, W1, BT, S15, H, W1, BT, W1, BT, W2,$ $S15, H, W13, H >$
User8	$< 4/S, 0/B, 5/H, 3/BT, 6/W >$
User9	$< S5, H, S12, W8, S12, W14, S12, W8, S12, W8 >$
User9	$< 5/S, 0/B, 1/H, 0/BT, 4/W >$
User19	$< S15, H, W20 >$
User19	$< 1/S, 0/B, 1/H, 0/BT, 1/W >$
User20	$< S15, H, S15, H, S6, W8 >$
User20	$< 3/S, 0/B, 2/H, 0/BT, 1/W >$

Table 6.16. Summary of Successful Users

Users(S)	S	B	BT	H	W
2	1	2	0	1	1
3	7	0	1	6	4
6	5	0	1	3	4
10	3	0	1	1	3
11	4	0	1	4	1
12	3	0	1	1	3
13	6	4	2	1	4
14	8	0	4	3	9
15	1	1	0	1	1
16	2	1	1	3	1
17	1	0	0	1	1
18	2	0	0	2	2
Max	8	2	4	6	9
Min	1	0	0	1	1
Ave	3.58	.67	1	2.25	2.83
Sum	43	8	12	26	34

which is has the correct answer but fails to recognize the answer in the selected Web page. If we label the query "Limbic function of the brain" as S_{15}, and the query "limbic system" as S_6 (the numbering being related to the order in which users were recorded and their queries indexed and labeled), and the Web page selected as W_8, then $user_{20}$ as a bag is:

Table 6.17. Summary of Unsuccessful Users

Users(F)	S	B	BT	H	W
1	4	0	4	3	7
4	10	0	0	7	5
5	3	1	2	5	4
7	2	1	0	3	1
8	4	0	3	5	6
9	5	0	0	1	4
19	1	0	0	1	1
20	3	0	0	2	1
Max	10	1	4	7	7
Min	1	0	0	1	1
Ave	4	0.25	1.13	3.38	3.63
Sum	32	2	9	27	29

$user_{20} = \{S_{15}, H, S_{15}, H, S_6, W_8\}$ or also as:
$user_{20} = \{2/S_{15}, 1/S_6, 2/H, 0/BT, 1/W_8\}$, or also as:
$user_{20} = \{3/S, 0/B, 2/H, 0/BT, 1/W\}$.

A set is a special kind of a bag. Thus any result that applies to bags can be applied to sets when the latter is viewed as a bag.

Def: A set SU is said to be a support for a bag A iff:

$$Count_A(x) > 0 \text{ if } x \in SU$$
$$Count_A(x) = 0 \text{ if } x \notin SU.$$

Example: Thus in the case of the bag $user_{20}$, a support set SU_{20} is equal to: $SU_{20} = \{S,H,W\}$.

Def: The intersection of 2 bags E and F is a new bag D [131] such that for any $x \in X$:

$$Count_D(x) = Min[Count_E(x), Count_F(x)] \tag{6.23}$$

Example on Successful Users: The intersection of all successful users is a new crisp bag or a new user that has the following characteristics:

$$Count_{DS}(x) = Min[\sum_{i=1}^{i=12} Count_{Ei}(x)]$$

$$= < 1/S, 0/B, 1/H, 0/BT, 1/W >$$

$Count_{DS}(x)$ corresponds to $user_{15}$ (see table 6.14).

Example on Unsuccessful Users: The intersection of all unsuccessful users is a new bag or new user that has the following characteristics:

$$Count_{DF}(x) = Min[\sum_{i=1}^{i=8} Count_{Ei}(x)]$$
$$= \ <1/S, 0/B, 1/H, 0/BT, 1/W>$$

$Count_{DF}(x)$ corresponds to $user_{19}$ (see table 6.15).
Also notice that $Count_{DS}(x) = Count_{DF}(x)$

Finding on "Intersection" of successful and unsuccessful bags: Both Successful and unsuccessful users have in common a crisp bag or a user that searched 1 time, did 1 hyperlink, and found the answer in 1 Web page or quit after 1 one Web page and failed to find an answer.

Def: The addition of two bags E and F drawn from X is a new bag C denoted C=A $\oplus$ B [86] such that for each $x \in X$:

$$Count_C(x) = Count_E(x) + Count_F(x) \qquad (6.24)$$

Example on Successful Users: The sum of all successful users results in a new crisp bag CS that represents the sum of all successful users:

$$Count_{CS}(x) = [\sum_{i=1}^{i=12} Count_{Ai}(x)]$$
$$= \ <43/S, 8/B, 26/H, 12/BT, 34/W>$$

Example on Unsuccessful Users: The sum of all unsuccessful users results in a new crisp bag CF that represents the sum of all unsuccessful users:

$$Count_{CF}(x) = [\sum_{i=1}^{i=8} Count_{Ai}(x)]$$
$$= \ <32/S, 2/B, 27/H, 9/BT, 29/W>$$

Def: Assume E and F are two bags drawn from X. The removal of bag F from bag E is a new bag R = E $\ominus$ F such that for any $x \in X$:

$$Count_R(x) = Max[Count_E(x) - Count_F(x), 0] \qquad (6.25)$$

Example: The removal of all unsuccessful users bag CF from the successful users bag CS is a new crisp bag or a new user that has the following characteristics:

$$Count_R(x) = Max[Count_{CS}(x) - Count_{CF}(x), 0]$$
$$= Max[<43/S, 8/B, 26/H, 12/BT, 34/W>$$
$$- <32/S, 2/B, 27/H, 9/BT, 29/W>, 0]$$
$$= \ <11/S, 6/B, 0/H, 3/BT, 5/W>$$

Finding on removal of unsuccessful users bag from successful users bags: Removing the sum of all unsuccessful users bag CF from the sum of all successful users bag CS results in a new crisp bag that has 11 searches, 6 browses, 3 backtrackings, and visited 5 Web pages.

Applying Crisp bags to the "uni-variable" query

The uni-variable query that we consider is: **"Find the number of Web pages(W) of users whose number of searches(S) is less than average"**. To answer such a query, we consider the following definition.

Def: Selection from bags via set specification [93]: Let A be a bag drawn from the set X and let $Q \subset X$. We are interested in forming a new bag, which consists of the elements of bag A that have membership in the set Q. This new bag denoted by $D = A \otimes Q$ where the count function for D is:

$$Count_D(x) = Count_A(x) \text{ if } x \in Q \qquad (6.26)$$
$$Count_D(x) = 0 \text{ if } x \notin Q$$

Example on Successful users:

We consider the set Q={S,W} $\subset$ X={S,B,H,BT,W}. Applying 6.26 to Q and X will result in a bag DS for successful users:
$Count_{DS}(x) = << 1S,1W >,< 7S,4W >,< 5S,4W >,< 3S,3W >,<$
$4S,1W >,< 3S,3W >,< 6S,4W >,< 8S,9W >,< 1S,1W >,< 2S,1W >,$
$< 1S,1W >,< 2S,2W >>$ By looking at the elements that belong to $Count_{DS}(x)$, for example, the element $< 7S,4W >$ has a number of searches equal to 7 and a number of Web pages equal to 4. Let α be the average of the number of searches for the successful users which is equal to 3.58 as seen in table 6.16. The element $< 7S,4W >$ should not be part of the answer to the query since for $S = 7 > 3.58$. There is no such an "operation" in the theory of Crisp bags as developed in ([131, 132]) that can select elements from a crisp bag according to a given threshold.

Example on Unsuccessful users: Applying 6.26 to Q and X will result in a bag DF for unsuccessful users:
$Count_{DF(x)} = << 4S,7W >,< 10S,5W >,< 3S,4W >,< 2S,1W >,$
$< 4S,6W >,< 5S,4W >,< 1S,1W >,< 3S,1W >>$

Summary of Applying Crisp Bags to the "Uni-variable" Query

By looking at the elements that belong to $Count_{DF}(x)$, for example, the element $< 10S,5W >$ has a number of searches equal to 10 and a number of Web pages equal to 5. Let α be the average of the number of searches for the unsuccessful users which is equal to 4 as seen in table 6.17. The element $< 10S,5W >$ should not be part of the answer to the query since for $S = 10 > 4$. There is no such an "operation" in the theory of Crisp bags as developed in ([131, 132]) that can select elements from a crisp bag according to a given threshold. Thus, Crisp bags with the set of operators in ([131, 132]) failed to answer the query "Find the number of Web pages (W) of users whose number of searches(S) is less than the average". Fuzzy bags are considered next to help answer such a query.

6.4.3 One Dimensional "Fuzzy Bag" Modeling User's Actions

Def: A fuzzy bag on a set X [92] is characterized by a characteristic function ψ which associates to every x $\in$ X a bag of degrees:

$$\psi_A : X \to Bag \qquad (6.27)$$

Notation: In a fuzzy bag, as we do in a fuzzy sets, we separate the elements from their characteristic values by the symbol //. Thus, a fuzzy bag, made out of the elements of $X = \{x_1, x_2, x_3, \cdots, x_n\}$ having respectively $\{\psi_1, \psi_2, \cdots, \psi_n\}$ characteristic values is represented by:

$$\{\psi_1//x_1, \psi_2//x_2, \psi_3//x_3, \cdots \psi_n//x_n\} \qquad (6.27')$$

(6.27') yields a fuzzy bag that can be represented by a simplified version:

$$A = \{< a_1//x_1, a_2//x_2, a_3//x_3, \cdots, a_n//x_n >\} \text{ where } a_i = \psi_i \quad (6.28)$$

In the case of our user's behavior, the set X = s{S,B,H,BT,W} will take on fuzzy attributes. The "uni-variable" query in the abstract can be answered from a fuzzy bag perspective.

Applying One Dimensional Fuzzy Bags to the "Uni-variable" Query

Example on Successful users: By applying ψ_A to successful users, it looks like ψ_{AS}:

$\psi_{AS} = \{1W \to << .6, .8, .9, .9, .9 >>, 2W \to << .8 >>, 3W \to << .7, .7 >>, 4W \to << .3, .4, .5 >>, 9W \to << .2 >>\}$

Thus, according to 6.28, the fuzzy bag ψ_{AS} for successful users is made out of elements (average =3.58, see table 6.16):

$\psi_{AS} = \{<< .6, .8, .9, .9, .9 >> //1W, << .8 > //2W, << .7, .7 >> //3W, << .3, .4, .5 >> //4W, << .2 >> //9W\}$.

Figure 6.13 represents such a fuzzy bag. Note that the first 5 circles have diameters of bubble size of 1W, the next circle is a bigger since it is for 2W, the next circles are for 3W, the next ones are for 4 W, and the last big one is for 9W. Note also that the values showed in the circles are the bubble sizes.

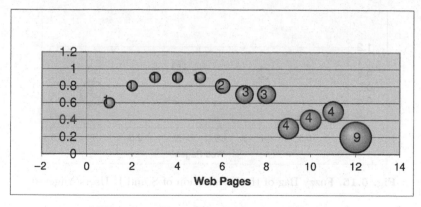

Fig. 6.13. Fuzzy Bag for Successful Users

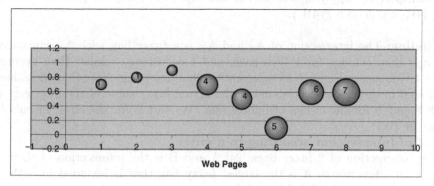

Fig. 6.14. Fuzzy Bag for Unsuccessful Users

Example on Unsuccessful users: By applying ψ_A to unsuccessful users, it looks like ψ_{AF}: $\psi_{AF} = \{1W \rightarrow << .7, .8, .9 >>, 4W \rightarrow << .7, .5 >>, 5W \rightarrow << .1 >>, 6W \rightarrow << .6 >>, 7W \rightarrow << .6 >> .\}$

Thus, according to 6.28, the fuzzy bag ψ_{AF} for unsuccessful users is made out of elements (average=4, see table 6.17):
$\psi_{AF} = \{ << .7, .8, .9 >> //1W, << .7, .5 >> //4W, << .1 >> //5W, << .6 >> //6W, << .6 >> //7W\}$.

Figure 6.14 represents such a fuzzy bag. Note that the first 3 circles have diameters of bubble size of 1W, the next 2 circles are for 4W, the next circle is for 5W, the next one is for 6W, and the last one is for 7W.

intersection of 2 fuzzy bags

Def: The intersection of 2 fuzzy bags ([131, 132]) A and B is the largest fuzzy bag that is included in both A and B. The number of occurrences of the elements of intersection is the minimum value of the number of occurrences of x in A and the number of occurrences of x in B:

$$A \cap B = \{wx | min(w_A(x), w_B(x))\} \qquad (6.29)$$

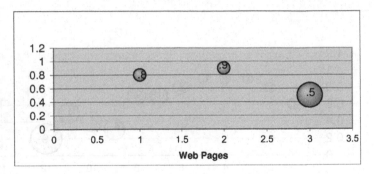

Fig. 6.15. Fuzzy Bag of the Intersection of S and U Users(Yager's)

Example: By applying 6.29 to A_S and A_F : $A_S \cap A_F = \{<< .8, .9 >> //1W, << .5 >> //4W\}$.

Finding: The intersection of A_S and A_F is a fuzzy bag with 2 occurrences larger than average for 1 Web page, and 1 occurrence smaller than average for 4 Web pages. Figure 6.15 represents such a fuzzy bag. Note that the values showed in the circles correspond to the actual fuzzy value of the intersection of .8 and .9 for the circles of bubble size 1W, and .5 for the circle of bubble size of 4W.

Def of Intersection according to α cuts:
The intersection of 2 fuzzy bags [24] A and B is the intersection of their α cuts. In other words, it is the largest fuzzy bag that is included in both A and B. To apply the definition of the α cuts we sort the elements x of A and x of B in decreasing order and then to apply a "min" operator to an element x of A and an element x of B:

$$\forall \alpha : (A_\alpha \cap B_\alpha) = (A_\alpha \cap B_\alpha) \tag{6.30}$$

Example: Applying 6.30 to A_S and A_F yields: $(A_S \cap A_F) = \{ << .7, .8, .9 >> //1W, << .5 >> //4W\}$.

Finding: The intersection of A_S and A_F is a new fuzzy bag with 3 occurrences at a degree above average for 1 Web page, and 1 occurrence at a degree smaller than average for 4 Web pages. Figure 6.16 represents such a fuzzy bag. The values showed in the circles correspond to the actual fuzzy value of the intersection of .7, .8 and .9 for the circles of bubble size 1W, and .5 for the circle of bubble size of 4W.

difference of 2 fuzzy bags

Def: The difference of 2 fuzzy bags ([131, 132]) A and B is the smallest fuzzy bag S such that the addition of S and B contains A. In other words, the difference of 2 fuzzy bags is a fuzzy bag that contains the elements of A and does not contain the elements of B. The number of occurrences of the elements of union is the maximum value of the difference of the number of occurrences of x in A and the number of occurrences of x in B:

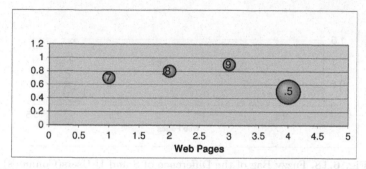

Fig. 6.16. Fuzzy Bag of the Intersection of S and U Users(Connan's)

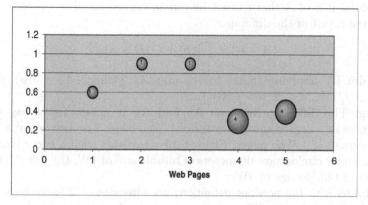

Fig. 6.17. Fuzzy Bag of the Difference of S and U Users(Yager's)

$$A - B = \{w * x | max(0, w_A(x) - w_B(x))\} \qquad (6.31)$$

Example: By applying (6.31) to A_S and A_F yields: $(A_S - A_F) = \{<<$ .9, .9, .6 $>> //1W, << .4, .3 >> //4W\}$.

Finding: *The difference between* A_S *and* A_F is a new fuzzy bag with 3 occurrences among which 2 occurrences are at a very high degree and 1 occurrence at a degree below average of 1 Web page, and 2 occurrences at a degree below average of 4 Web pages. Figure 6.17 represents such a fuzzy bag. Note that the first 3 circles have diameters of bubble size of 1W, the last 2 circles have diameters of bubble size of 4W.

Another Def of Difference: The difference of 2 fuzzy bags [24] A and B is based on the additive nature of the difference of "A-B". It is based on the idea of what fuzzy bag "S" to be added to "B" to get a fuzzy bag that contains the fuzzy bag "A". To apply such a definition, we sort the elements x of A and x of B in decreasing order. Every occurrence of an element x of A is matched with an element x of B which is greater or equal to that x of A.

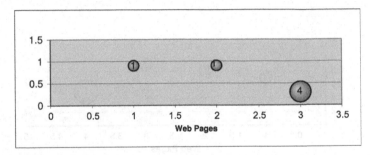

Fig. 6.18. Fuzzy Bag of the Difference of S and U Users(Connan's)

Every element x of A that cannot be put in correspondence to element x of B is in the result of the difference:

$$(A - B) = \bigcap \{S | A \subseteq (B + S)\} \tag{6.32}$$

Example: By applying (6.32) to A_S and A_F yields: $(A_S - A_F) = \{<< .9, .9 >> //1W, << .3 >> //4W\}$.

Finding: The difference between A_S and A_F is a new fuzzy bag with 2 occurrences at a degree above average of 1 Web page, 1 occurrence at a degree below average of 4 Web pages. Figure 6.18 represents such a fuzzy bag. Note that the first 2 circles have diameters of bubble size of 1W, the last circle have diameter of bubble size of 4W.

Problems with the previous definitions on difference of 2 fuzzy bags In the classical theory, the set difference can be defined without complementation by: $A - B = S \Longleftrightarrow A = (A \cap B) \cup S$. In bag theory, this definition has an additive semantics. A - B is associated with the elements, which have to be added to (A ∩ B) so that it equals:

$$A : A - B = S \Longleftrightarrow A = (A \cap B) + S \tag{6.33}$$

Unfortunately, in the context of fuzzy bags, it is not always possible to find S, the relative complement of (A ∩ B) with respect to A. Let us consider the 2 fuzzy bags $A_S = \{<< .9, .9, .9, .8, .6 >> //1W\}$ and $A_F = \{<< .9, .3 . > /1W\}$.

$(A_S \cap A_F) = \{<< .9, .3 >> //1W\} = A_F$ and there is no S such that $A_S = A_F + S$ because the occurrence of 1W with the degree of 0.3 cannot be upgraded to a higher levels thanks to the addition.

What if fuzzy cardinalities ([10]) were used to define fuzzy numbers of occurrences of fuzzy bags?

Def of Fuzzy cardinality: The concept of fuzzy cardinality of a fuzzy set called FGCount(A) is defined by:

$$\forall n : \mu_{|A|}(n) = sup\{\alpha | |A| \geq n\} \tag{6.34}$$

Then considering this notion of fuzzy cardinality, the occurrences of an element x in a fuzzy bag A can be characterized as a fuzzy integer denoted by $\omega_A(x)$. This fuzzy number is the fuzzy cardinality of the fuzzy set of occurrences of x in A.

Def of fuzzy Bag according to Fuzzy cardinality:

Thus a fuzzy bag A, on a universe U, can be defined by the characteristic function Ω_A from U to N_f, where N_f is the set of fuzzy integers:

$$\Omega_A : U \to N_f \tag{6.35}$$

If $A_S = \{< .6, .8, .9, .9, .9 > //1W, < .8 > //2W, < .7, .7 > //3W,$ $< .3, .4, .5 > //4W, < .2 > //9W\}$, then:

$A_S(1W) = \{< 1 > /0, < .9 > /1, < .9 > /2, < .9 > /3, < 0.8 > /4, < .6 >$ $/5\}$, $A_S(2W) = \{< 1 > /0, < .8 > /1\}$, $A_S(3W) = \{< 1 > /0, < .7 >$ $/1, < .7 > /2\}$, $A_S(4W) = \{< 1 > /0, < .5 > /1, < .4 > /2, < .3 > /3\}$, $A_S(9W) = \{< 1 > /0, < 2 > /1\}$.

If $A_F = \{< .7, .8, .9 > //1W, < .7, .5 > //4W, < .1 > //5W, < .6 >$ $//6W, < .6 > //7W\}$, then: $A_F(1W) = \{< 1 > /0, < .9 > /1, < .8 > /2,$ $< .7 > /3\}$, $A_F(4W) = \{< 1 > /0, < .7 > /1, < .5 > /2\}$,

$A_F(5W) = \{< 1 > /0, < .1 > /1\}$, $A_F(6W) = \{< 1 > /0, < .6 > /1\}$, $A_F(7W) = \{< 1 > /0, < .6 > /1\}$,

Operators on fuzzy bags based on Fuzzy Cardinalities:

$$\Omega_{A \cup B} = min(\Omega_A(x), \Omega_B(x)) \tag{6.36}$$
$$\Omega_{A \cap B} = max(\Omega_A(x), \Omega_B(x)) \tag{6.37}$$
$$\Omega_{A-B} = (\Omega_A(x) + \Omega_B(x)) \tag{6.38}$$

How does fuzzy cardinalities [93] deal with the difference of 2 fuzzy numbers. **Def of Difference of two fuzzy bags based on fuzzy cardinalities:** The condition in 6.33 has to be relaxed. The difference of 2 fuzzy bags [93] A and B is based on a fuzzy bag S_1 which approximates S such that it is the greatest fuzzy bag S_i that is contained in A:

$$S_1 = \bigcup\{S_i : A \supseteq (A \cap B) + S_i\} \tag{6.39}$$

Example:

$$\Omega_{AS}(1W) \cap \Omega_{AF}(1W) = min\{\{< 1 > /0, < .9 > /1, < .9 > /2, < .9 > /3,$$
$$< 0.8 > /4, < .6 > /5\}, \{< 1 > /0, < .9 > /1,$$
$$< .8 > /2, < .7 > /3\}\}$$
$$= \{< 1 > /0, < .9 > /1, < .8 > /2, < .7 > /3\}$$
$$\Omega_{AS}(4W) \cap \Omega_{AF}(4W) = min\{\{< 1 > /0, < .5 > /1, < .4 > /2, < .3 > /3\},$$
$$\{< 1 > /0, < .7 > /1, < .5 > /2\}\}$$
$$= \{< 1 > /0, < .5 > /1, < .4 > /2\}$$
$$\Omega_{AS} \cap \Omega_{AF} = \{\{< .9, .8, .7 > /1W\}, \{< .5, .4 > /4W\}\}.$$

In other words: $\forall \alpha, \forall x : w(A \cap B)_\alpha(x) + wS_{1\alpha}(x) \leq wA_\alpha(x)$.
By applying 6.33 to A_S and A_F: $w(A_S \cap A_F)_\alpha(x) + wS_{1\alpha}(x) \leq wA_{S\alpha}(x)$.

$(A_S \cap A_F) = \{\{< 1 > /0, < 0.9 > /1, < .8 > /2, < .7 > /3 >\}1W,$
$\quad \{< 1 > /0, < .5 > /1, < .4 > /2\}4W\},$
$A_S(1W) = \{ << 1 >> /0, << .9 >> /1, << .9 >> /2, << .9 >> /3,$
$\quad << 0.8 >> /4, << .6 >> /5\}$
$A_F(1W) = \{ << 1/0 >>, << .9/1 >>, << .8/2 >>, << .7/3 >>\},$
$(A_S \cap A_F)(1W) = A_F(1W)$
$S_1(1W) = A_S(1W) - A_F(1W)$
$= \{< 1 > /0, < .9 > /1, < .9 > /2, < .9 > /3, < .8 > /4, < .6 > /5\}$
$-\{< 1 > /0, < .9 > /1, < .8 > /2, < .7 > /3\}$
$= \{< 1, 1 > /0, < .9 > /1, < .6 > /2\}.$

If $S_1(1W)$ is added to $A_F(1W)$, it is the closest to $A_S(1W)$:

$A_F(1W) + S_1(1W) = \{< 1 > /0, < .9 > /1, < .9 > /2, < .8 > /3, < .7 > /4,$
$< .6 > /5\} \subset \{< 1 > /0, < .9 > /1, < .9 > /2, < .9 > /3, < .8 > /4, < .6 > /5\}$
$\Omega_{AS}(4W) \cap \Omega_{AF}(4W) = \min\{\{< 1 > /0, < .5 > /1, < .4 > /2, < .3 > /3\},$
$\{< 1 > /0, < .7 > /1, < .5 > /2\}\} = \{< 1 > /0, < .5 > /1, < .4 > /2\}.$
$\Omega_{AS}(4W) \cap \Omega_{AF}(4W) \neq \Omega_{AF}(4W) = \{< 1 > /0, < .7 > /1, < .5 > /2\}.$

Thus the difference of 2 fuzzy bags A_S and A_F according to cardinality (see table 6.11) is:

$(A_S - A_F) = \{< .9, .6 > //1W, < .3 > //4W\}$

Finding: The difference between A_S and A_F is a new fuzzy bag with 2 occurrences where 1 is a degree at a degree above average and 1 is a degree below average for 1 Web page, and 1 occurrence at a degree below average for 4 Web pages. Figure 6.19 represents such a fuzzy bag. The first 2 circles have diameters of bubble size of 1W, the last circle have diameter of bubble size of 4W. The second circle of size 1W has a fuzzy value of .6 and not .9 as in Figure 6.19.

Fuzzy distance between 2 fuzzy bags

What if the operators of intersection and difference were not enough of an indication on how different or similar successful and unsuccessful users are? Can we establish a measure of similarity between 2 fuzzy bags?

Def: Let FA and FB are 2 fuzzy bags drawn from a set X. A measure of similarity between FA and FB denoted by S (FA, FB) is defined by:

$$S(FA, FB) = (1 - (1/Card(X)) \sum_{x \in X} (d(FA(x), FB(x)) \qquad (6.40)$$

Where d(FA(x), FB(x)) is defined by the α cuts in FA and FB:

$$d(FA(x), FB(x)) = \left(\frac{\sum \alpha FA(x)_\alpha}{\sum FA(x)_\alpha}\right) - \left(\frac{\sum \alpha FB(x)_\alpha}{\sum FB(x)_\alpha}\right) \qquad (6.41)$$

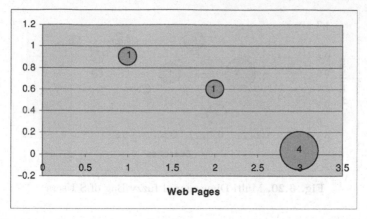

Fig. 6.19. Fuzzy Bag of the Difference of S and U Users(Cardinality)

Clearly, $0 \geq S(FA, FB) \leq 1$.

Properties of S:

- S is reflexive since S(FA,FA) = 1. Thus according to S, every fuzzy bag is similar to itself.
- S is symmetric since S(FA,FB) = S(FB,FA) because d(FA,FB)=d(FB,FA).
- S is not transitive since FA,FB, and FC according to (6.41): d(FA,FC) $\neq$ d(FA,FB)+d(FB,FC)

Example: By applying 6.41 to A_S and A_F yields:
$d(A_S(1W), A_F(1W)) = .011; \ d(A_S(4W), A_F(4W)) = .20.$
Applying (6.40) yields: $S(A_S, A_F)$=0.8945.

6.4.4 Multidimensional Fuzzy Bags modeling of User's Actions ([93])

To answer: "Find the number of Web pages of users whose number of searches is less than average and hyperlink navigation is less than the average", a multi-dimensional fuzzy bag is considered.

The Association of n fuzzy bags

$$(Att, Bag(X_1)), (Att, Bag(X_2)), \cdots, (Att, Bag(X_n)) \rightarrow$$
$$Bag(Tuple(Att : X_1, Att : X_2, \cdots, Att : X_n)) \qquad (6.42)$$

In the case of 2 fuzzy bags:

$$(X_1, Bag(X_1)), (X_2, Bag(X_2) \rightarrow Bag(Tuple(X_1, X_2)) \qquad (6.43)$$

The Select operator [93] applies a "fuzzy predicate p" (less than average for example) to each occurrence of the fuzzy bag in 6.43 and results in a new fuzzy bag:

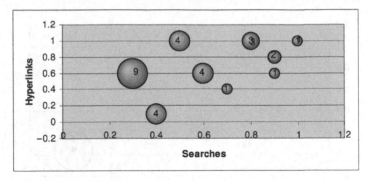

Fig. 6.20. Multi Dimesnional fuzzy Bag of S Users

$$Select(Bag, p) : Bag(Tuple(X_1, X_2), (X_1 \rightarrow MBag), \qquad (6.44)$$
$$(X_2 \rightarrow MBag)) \rightarrow Bag(Tuple(X_1, X_2))$$

where Mbag is a multi-dimensional bag.

Example on Successful Users: By applying (6.43) to successful users (see table 6.16) yields:

$\psi_{(S,H)}$ = $1W \rightarrow<< 1S, 1H >, < 1S, 1H, < 1S, 1H >, < 2S, 3H >, < 4S, 4H >>, 2W \rightarrow<< 2S, 2H >>, 3W \rightarrow< 2* < 3S, 1H >>, 4W \rightarrow<< 7S, 6H >, < 5S, 3H >, < 6S, 1H >>, 9W \rightarrow<< 8S, 3H >>$

By applying 6.45 to 6.43 and rewriting it in the form of 6.28, the fuzzy bag A_S for successful users is made out of elements:

A_S = $\{<< 1, 1 >, < 1, 1 >, < 1, 1 >, < .9, .6 >, < .7, .4 >> //1W, < .9, .8 > //2W, << .8, 1 >, < .8, 1 > //3W, << .4, .1 >, < .6, .6 >, < .5, 1 >> //4W, < .3, .6 > //9W\}$.

Figure 6.20 represents such a 2-D fuzzy bag. We have 3 circles at the same position S=1 and H=1 for W=1 and appear as one circle in Figure 6.20. Also we have 2 circles for W=3 at the same position S=.8 and H=.1 and appear as one circle in Figure 6.20. Thus, only 9 circles are seen in Figure 6.20 instead of 12 circles.

Example on Unsuccessful Users: By applying (6.43) to unsuccessful users (see table 6.17) yields:

$\psi_{(S,H)}$ = $1W \rightarrow<< 2S, 3H >, < 1S, 1H >, < 3S, 2H >>, 4W \rightarrow<< 3S, 5H >, < 5S, 1H >>, 5W \rightarrow<< 10S, 7H >>, 6W \rightarrow<< 4S, 5H >>, 7W \rightarrow<< 4S, 4H >>.$

By applying 6.45 to 6.43 and rewriting it in the form of 6.28, the fuzzy bag A_F for unsuccessful users is made out of elements:

$A_F = \{<< .9, .8 >, < 1, 1 >, < .8, .9 >> //1W, << .8, .6 >, < .6, 1 >> //4W, << .1, .4 >> //5W, << .7, .6 >> //6W, << .7, .7 >> //7W\}.$

Figure 6.21 represents such a fuzzy bag.

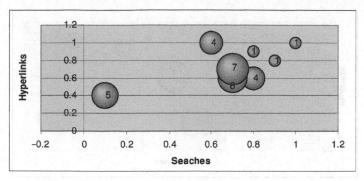

Fig. 6.21. Multi Dimesnional fuzzy Bag of F Users

These 2 degrees one for searches, and one for the hyperlinks, are commensurable, for example a degree 0.5 from the first predicate and a degree 0.5 from the second predicate do not have the same meaning. Consequently they cannot be combined, and as such any of the 2 dimensions cannot be reduced. Thus, the fuzzy bag has been coined a multi-dimensional fuzzy bag.

Intersection of two multidimensional fuzzy bags

Def: The intersection of 2 fuzzy bags A and B [131, 132] is the largest fuzzy bag that is included in both A and B. The number of occurrences of the elements of intersection is the minimum value of the number of occurrences of x in A and the number of occurrences of x in B:

$$A \cap B = \{wx | min(w_A(x), w_B(x))\} \qquad (6.45)$$

Example: By applying (6.45) to A_S and $A_F : A_S \cap A_F = \{< 1, 1 > //1W\}$

Finding: The intersection of A_S and A_F is a 2 dimensional fuzzy bag with 1 occurrence made out of an element larger than average in the number of searches and larger than average in the number of hyperlink navigation for 1 Web page (both bags being equally dominant).

Def of Intersection according to α cuts:

The intersection of 2 fuzzy bags A and B is the intersection of their α cuts. In other words, it is the largest fuzzy bag that is included in both A and B. To apply the definition of the α cuts we sort the elements x of A and x of B in decreasing order and then to apply a "min" operator to an element x of A and an element x of B:

$$\forall \alpha : (A \cap B)_\alpha = (A_\alpha) \cap (B_\alpha) \qquad (6.46)$$

Example: By applying (6.46) to A_S and A_F:

$(A_S \cap A_F) = \{<< 1, 1 >, < .9, .8 >, < .8, .9 >> //1W, << .6, .6 >, < .5, 1 >> //4W\}$

Finding: the intersection of A_S and A_F is a new fuzzy bag with 3 occurrences among which 1 occurrence is made out of an element larger than average in the number of searches and larger than average in the number of

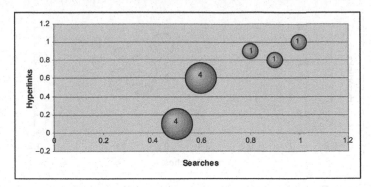

Fig. 6.22. Multi Dimensional fuzzy Bag of the intersection of S and U users(α cuts)

hyperlinks (common to both bags) and 2 occurrences made out of an element larger than average in the number of searches and larger than average in the number of hyperlinks (coming from the failure bag) for 1 Web page; and 2 occurrences where 1 occurrence is made out of an element equal to average in the number of searches and equal to average in the number of hyperlinks for 4 Web pages(coming from the success bag) and 1 occurrence made out of an element smaller than average in the number of searches and smaller than average in the number of hyperlinks for 4 Web pages(coming from the success bag). Figure 6.22 represents the fuzzy bag of the intersection.

Def of Intersection according to the intersection of fuzzy numbers: The intersection of 2 fuzzy bags A and B according to ([92]) is the intersection of their fuzzy numbers. In other words, it is the largest fuzzy number that is included in both A and B. To apply the definition of ([93]), we sort the elements x of A and x of B in decreasing order and then to apply a "min" operator to an element x of A and an element x of B.

Example: By applying ([93]) to A_S and A_F yields:

$\Omega_{AS \cap AF}(1W) = min(\Omega_{AS}(1W), \Omega_{AF}(1W))$

$\Omega_{AS \cap AF}(1W) = min(\{< 1, 1 > /0, < 1, 1 > /1, < 1, 1 > /2, < 1, 1 > /3, < 0.9, .6 > /4, < .7, .4 > /5\}, \{< 1, 1 > /0, < 1, 1 > /1, < .9, .8 > /2, < .8, .9 > /3\}) = \{< 1, 1 > /0, < 1, 1 > /1, < .9, .8 > /2, < .8, .9 > /3\}$

$\Omega_{AS \cap AF}(4W) = min(\Omega_{AS}(4W), \Omega_{AF}(4W))$

$\Omega_{AS \cap AF}(4W) = min(\{< 1, 1 > /0, < .6, .6 > /1, < .5, .1 > 2, < .4, .1 > /3\}, \{< 1, 1 > /0, < .8, .6 > /1, < .6, .1 > /2\}) = \{< 1, 1 > /0, < .6, .6 > /1, < .5, .1 > /2\}$

$A_S \cap A_F = \{\{1, 1 > /0, < 1, 1 > /1, < .9, .8 > /2, < .8, .9 > /3\}1W, \{< 1, 1 > /0, < .6, .6 > /1, < .5, .1 > /2\}4W\}.$

Finding: This is the same result obtained using α cuts.

Difference of two Multidimensional fuzzy bags

Def: Yager's: ([131, 132]) difference of 2 fuzzy bags A and B is the smallest fuzzy bag S such that the addition of S and B contains A. In other words, the difference of 2 fuzzy bags is a fuzzy bag that contains the elements of

A and does not contain the elements of B. The number of occurrences of the elements of the difference is the maximum value of the difference of the number of occurrences of x in A and the number of occurrences of x in B:

$$A - B = \{w * x | max(0, w_A(x) - w_B(x))\} \quad (6.47)$$

Example: Applying 6.47 to A_S and A_F yields:

$A_S - A_F = \{< max(0, 3 < 1, 1 > -1 < 1, 1 >), max(0, 1 < .9, .6 > -0$
$< .9, .6 >), max(0, 1 < .7, .4 > -0 < .7, .4 >)//1W, < max(0, 1 < .6, .6 >$
$-0 < .6, .6 >)//4W, max(0, 1 < .5, .1 > -0 < .5, .1 >)//4W, max(0, 1 <$
$.4, .1 > -0.4, .1 >) > //4W\}$
$A_S - A_F = \{< 2 < 1, 1 >, < .9, .6 >, < .7, .4 >> //1W, << .6, .6 >, < .5, .1 >$
$, < .4, .1 >> //4W\}$
$A_S - A_F = \{<< 1, 1 >, < 1, 1 >, < .9, .6 >, < .7, .4 >> //1W, << .6, .6 >,$
$< .5, 1 >, < .4, .1 >> //4W\}$

Finding: The difference between A_S and A_F is a new fuzzy bag with 4 occurrences where 2 are made out of an element larger than average in the number of searches and larger than average in the number of hyperlinks, 1 occurrence made out of an element larger than average in the number of searches and smaller than average in the number of hyperlinks, and 1 occurrence made out of an element equal to average in the number of searches and smaller than average in the number of hyperlinks for 1 Web page; and 3 occurrences where 1 occurrence is made out of an element equal to average in the number of searches and equal to average in the number of hyperlinks, and 2 occurrences made out of an element smaller than average in the number of searches and smaller than average in number of hyperlinks for 4 Web pages. Figure 6.23 represents such a fuzzy bag. Thus in this case 2 circles are at the same position S=1 and H=1 for W=1 and appear as one circle in Figure 6.23.

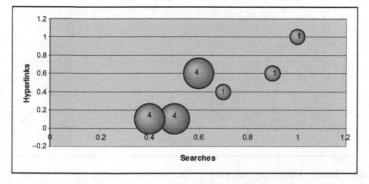

Fig. 6.23. Multi Dimesnional fuzzy Bag of the difference between S and F Users(Yager's)

Def: Connan's ([24]) difference of 2 fuzzy bags A and B is based on the idea of what fuzzy bag "S" to be added to "B" to get a fuzzy bag that contains the fuzzy bag "A". To apply such a definition, we sort the elements of A and B in decreasing order. Every occurrence of an element of A is matched with an element of B which is greater or equal to that of A. Every element of A that cannot be put in correspondence to element of B is in the result of the difference:

$$A - B = \bigcap\{S | A \subseteq (B + S)\} \tag{6.48}$$

Example: Applying (6.48) to A_S and A_F yields: $(A_S - A_F) = \{<< 1, 1 >, < 1, 1 >> //1W, < .4, .1 > //4W\}$

Finding: The difference between A_S and A_F is a new fuzzy bag with 2 occurrences made out of an element larger than average in the number of searches and larger than average in the number of hyperlink navigations for 1 Web page, and 1 occurrence is made out of an element smaller than average in the number of searches, and smaller than average in the number of hyperlink navigations for 4 Web pages. Figure 6.24 represents such a difference. Thus in this case 2 circles are at the same position S=1 and H=1 for W=1 and appear as one circle in Figure 6.24.

Def of Difference based on Fuzzy cardinalities: How does fuzzy cardinalities [93] deal with the difference of 2 fuzzy numbers in the case of a 2 dimensional fuzzy bag.

Def: As stated above, the condition in (6.33) has to be relaxed. Thus a fuzzy bag S_1 which approximates S such that it is the greatest fuzzy bag S_i that is contained in A:

$$S_1 = \bigcup\{S_i : A \supseteq (A \cap B) + S_i\} \tag{6.49}$$

In other words, $\forall \alpha$, $\forall x$: $w_{A \cap B}(x) + w_{S1\alpha}(x) \leq w_{A\alpha}(x)$.

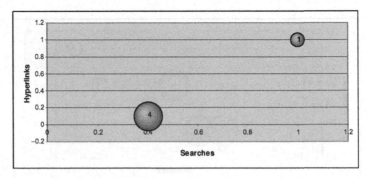

Fig. 6.24. Multi Dimesnional fuzzy Bag of the difference between S and F Users(Connan's):

Example: Let us apply (6.49) to A_S and A_F:

$w_{A_S \cap A_F} \alpha(x) + w_{S1}(x) \leq w A_S(x)$ to A_S and A_F

$A_S \cap A_F = \{\{1, 1 > /0, < 1, 1 > /1, < .9, .8 > /2, < .8, .9 > /3\} * 1W,$
$\qquad\qquad \{< 1, 1 > /0, < .6, .6 > /1, < .5, .1 > /2\} * 4W\}\}$

$A_S(1W) = \{<< 1, 1 >, < 1, 1 >, < 1, 1 >, < .9, .6 >, < .7, .4 >>\}$

$A_S \cap A_F = \{<< 1, 1 >, < 1, 1 >, < 1, 1 >, < .9, .6 >, < .7, .4 >> //1W\},$

$A_F(1W) = \{<< 1, 1 >, < .9, .8 >< .8, .9 >> //1W\}$

$(A_S - A_F)(1W) = A_F(1W)$

Extending the results to fuzzy numbers requires the following calculations:
$A_S \cap A_F = \{< 1, 1 > /0, < 1, 1 > /1, < .9, .8 > /2, < .8, .9 > /3 > //1W,$
$< 1, 1 > /0, < .6, .6 > /1, < .5, .1 > /2//4W\}$

$A_S(1W) = \{< 1, 1 >, < 1, 1 >, < 1, 1 >, < .9/.6 >, < .7, .4 >\} \rightarrow$
$\Omega_{A_S}(1W) = \{< 1, 1 > /0, < 1, 1 > /1, < 1, 1 > /2, < 1, 1 > /3, < .9/.6 >$
$/4, < .7, .4 > /5\},$
$\Omega_{A_F}(1W) = \{< 1, 1 > /0, < 1, 1 > /1, < .9, .8 > /2, < .8, .9 > /3\}$
$\Omega_{A_S \cap A_F}(1W) = \{< 1, 1 > /0, < 1, 1 > /1, < .9, .8 > /2, < .8, .9 > /3\}$
By extending the results to bags, we could say:

$$\forall x : \Omega_{A_S}(x) = \Omega_{A_S \cap A_F}(x) + \Omega_S(x) \qquad (6.50)$$

We know that such solution (6.50) does not exist. The best we can hope for is:

$$\forall x : \Omega_{A_S \cap A_F}(x) + \Omega_{S1}(x) \leq \Omega_{A_S}(x) \qquad (6.51)$$

$S_1(1W) = A_S(1W) - A_F(1W)$
$\qquad = \{< 1, 1 > /0, < 1, 1 > /1, < 1, 1 > /2, < 1, 1 > /3,$
$\qquad\quad < .9/.6 > /4, < .7, .4 > /5\} - \{< 1, 1 > /0, < 1, 1 > /1,$
$\qquad\quad < .9, .8 > /2, < .8, .9 > /3\}$
$\qquad = \{< 1, 1 > /0, < 1, 1 > /1, < .7, .4 > /2\}.$

If $S_1(1W)$ is added to $A_F(1W)$, it is the closest to $A_S(1W)$:

$A_F(1W) + S_1(1W) = \{< 1, 1 > /0, < 1, 1 > /1, < 1, 1 > /2, < .9, .8 > /3,$
$< .8, .9 > /4, < .7, .4 > /5\} \subseteq \{< 1, 1 > /0, < 1, 1 > /1, < 1, 1 > /2,$
$< 1, 1 > /3, < .9/.6 > /4, < .7, .4 > /5\}$
$(A_S \cap A_F) = \{< 1, 1 > /0, < 1, 1 >, < .9, .8 >, < .8, .9 > //1W, < .6, .6 >,$
$< .5, .1 > //4W\},$

$(A_S \cap A_F)(4W) = \{< .6, .6 > /1, < .5, .1 >\} \rightarrow$

$\Omega_{A_S \cap A_F}(4w) = \{< 1, 1 > /0, < .6, .6 > /1, < .5, .1 > /2\},$

$\Omega_{A_F}(4W) = \{< 1, 1 > /0, < .8, .6 > /1, < .6, .1 > /2 \neq \Omega_{A_S \cap A_F}(4W)$

$= \{< 1, 1 > /0, < .6, .6 > /1, < .5, .1 > /2\},$

$\Omega_{A_S}(4W) = \{< 1, 1 > /0, < .6, .6 > /1, < .5, .1 > /2, < .4, .1 > /3\},$

$\Omega_{A_S}(4W) - \Omega_{A_F}(4W) = \{< 1, 1 > /0, < .6, .6 > /1, < .5, .1 > /2, < .4, .1 > /3\}$

$-\{< 1, 1 > /0, < .8, .6 > /1, < .6, .1 > /2\}$

$= \{< 1, 1 > /0, < .4, .1 > /1\},$

$A_S - A_F = \{<< 1, 1 >, < .7, .4 >> //1W, << .4, .1 >> //4W\}$

(see tables 6.12 and 6.13)

Finding: The difference between A_S and A_F according to cardinality [93] is a new fuzzy bag with 2 occurrences among which 1 occurrence is at a degree above average in the number of searches and above average in the number of hyperlink navigations, and 1 occurrence is at a degree above average in the number of searches and below average in the number of hyperlink navigations for 1 Web page; and 1 occurrence at a degree below average in the number of searches, and below average in the number of hyperlink navigations for 4 Web pages. Figure 6.25 represents such a difference.

Def: Let FA and FB are 2 two dimensional fuzzy bags drawn from a set X. A measure of similarity between FA and FB denoted by S(FA, FB) is defined by:

$$S(FA, FB) = (1 - (1/Card(X)) \sum_{x \in X} (d(FA(x), FB(x))$$

where d(FA(x), FB(x)) is defined by the α cuts in FA and FB:

$$d(FA(x), FB(x)) = \frac{\sum < \alpha, \beta > FA(x)_{\alpha,\beta}}{\sum FA(x)_{\alpha,\beta}} - \frac{\sum < \alpha, \beta > FB(x)_{\alpha,\beta}}{\sum FB(x)_{\alpha,\beta}} \qquad (6.52)$$

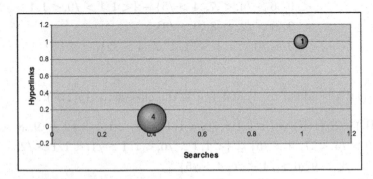

Fig. 6.25. Multi Dimesnional fuzzy Bag of the difference between S and F Users (α cuts:)

Clearly, $0 \leq S(FA, FB) \leq 1$ **Properties of S:** they are similar to the S for One dimensional fuzzy bags.

Example: By applying (6.52) to A_S and A_F yields:

$$d(A_S(1W), A_F(1W)) = .556; d(A_S(4W), A_F(4W)) = .23$$

Applying (6.40) yields: $S(A_S, A_F)=0.607$.

6.4.5 Discussion and Concluding Remarks

This study shows that fuzzy bags are a viable "computational tool" to interpret user's behavior. Although fuzzy bags since their creation by Yager have had few applications, more recently there is a resurgence of their application in multi-criteria decision making problems ([8]), foundation of multi-sets ([43]), and databases ([24]). Yager's theory on fuzzy bags has the disadvantage of not generalizing their operators on fuzzy sets. Connan's thesis on fuzzy bags [24] anchors fuzzy bags in Zadeh's definition of fuzzy sets [135] and α cuts, but failed to implement it in the case of the difference between 2 fuzzy bags, where Bosc and Rocacher[10] did which is compatible with the operations on fuzzy sets. Connan's interpretation of fuzzy bags [24], and Bosc and Rocacher [10], has helped enrich user's behavior and offered a more flexible interpretation of the collected data of user's behavior. This study further enhances the use of fuy bags to answer more questions about user's behavior. As such it becomes the only de facto technique when more complex queries are used. For example, if more than one attribute of the user's behavior was to be used, then better fuzzy bags operators will be needed to answer such a query. The results in figures 6.20, 6.21, 6.22, 6.23, 6.24, and 6.25 show a 2-dimensional bubble chart that were fuzzy on both attributes the number of searches and the number of hyperlink used. Compared to a study of how elementary school students ehavior on a task was done [62]:

> "Successful students looped searches and hyperlinks less than unsuccessful ones, examined a higher level of hyperlinks and homepages, and scrolled a lightly higher percentage of the screens and pages returned,"

This study showed a more complex comparison in nature and richer descriptive results between users who succeeded and those who failed through the use of queries. A major area of research is in the mining of user's behavior while searching the Web. This research has filled in a critical need of looking at the gathered data from users searching the Web in a totally different perspective and help answer queries that could have not been answered in a traditional setting. Most all studies on users behavior focuses on describing the characteristics of users behavior on searching, browsing, and hyperlink navigation and insist on calculating averages and standard deviation to describe the differences between users who succeeded and users who failed. We show that user's actions while searching the Web can be converted into a crisp bag but where the queries cannot be answered. By using [86, 92, 93], uni-variable queries can be answered and help into mining user's behavior and enrich the interpretation of such a behavior. fuzzy bags are the only viable "computational tool" to interpret user's behavior. Also, queries on comparing successful users to unsuccessful users while searching the Web in the new perspective lead to new applications of fuzzy bags.

The uni-variable query can be answered in the new perspective. The intersection of the 2 1-D fuzzy bags in figures 6.13 and 6.14 according:

- to ([131, 132]) is a fuzzy bag with 2 occurrences at a degree above average for 1 Web page, and 1 occurrence below average for 4 Web pages.
- to [24] is a fuzzy bag with 3 occurrences among which 2 are at a degree above average and 1 occurrence at a degree below average for 1 Web page, and 1 occurrence at a degree below average for 4 Web pages.

The difference of the 2 1-D fuzzy bags in figures 6.13 and 6.14 according:

- to ([131, 132]) is a fuzzy bag with 3 occurrences among which 2 occurrences are a degree above average for 1 Web page and 2 occurrences at a degree below average for 4 Web pages.
- to ([92, 93]) is a fuzzy bag with 2 occurrences that are above average for 1 Web page and 2 occurrences at a degree below average for 4 Web pages.
- to ([10]) is a fuzzy bag with 2 occurrences where 1 is a degree at a degree above average and 1 is a degree below average for 1 Web page, and 1 occurrence at a degree below average for 4 Web pages.

The multi-variable query can be answered in the new perspective after extending the case of one dimensional fuzzy bag to Multi-dimensional fuzzy bags. The intersection of the two dimensional fuzzy bags in figures 6.20 and 6.21 according:

- to ([131, 132]) is a two dimensional fuzzy bag with 1 occurrence made out of an element larger than average in the number of searches and larger than average in the number of hyperlink navigation for 1 Web page (both bags being equally dominant).
- to ([92, 93]) is a two dimensional fuzzy bag with 3 occurrences among which 1 occurrence is made out of an element larger than average in the number of searches and larger than average in the number of hyperlinks (common to both bags) and 2 occurrences made out of an element larger than average in the number of searches and larger than average in the number of hyperlinks (coming from the failure bag) for 1 Web page; and 2 occurrences where 1 occurrence is made out of an element equal to average in the number of searches and equal to average in the number of hyperlinks for 4 Web pages (coming from the success bag) and 1 occurrence made out of an element smaller than average in the number of searches and smaller than average in the number of hyperlinks for 4 Web pages (coming from the success bag).

The difference of the two dimensional fuzzy bags in figures 6.20 and 6.21 according:

- to ([131, 132]) is a two dimensional fuzzy bag with 4 occurrences where 2 are made out of an element larger than average in the number of searches and larger than average in the number of hyperlinks, 1 occurrence made out of an element larger than average in the number of searches and smaller than average in the number of hyperlinks, and 1 occurrence made out of an element equal to average in the number of searches and smaller than average in the number of hyperlinks for 1 Web page; and 3 occurrences where 1 occurrence is made out of an element equal to average in the number of searches and equal to average in the number of hyperlinks, and 2 occurrences made out of an element smaller than average in the number of searches and smaller than average in number of hyperlinks for 4 Web pages.
- to ([92, 93]) is a two dimensional fuzzy bag with 2 occurrences made out of an element larger than average in the number of searches and larger than average in the number of hyperlink navigations for 1 Web page, and 1 occurrence is made out of an element smaller than average in the number of searches, and smaller than average in the number of hyperlink navigations for 4 Web pages.
- to ([86]) is a two dimensional fuzzy bag with 2 occurrences among which 1 occurrence is at a degree above average in the number of searches and above average in the number of hyperlink navigations, and 1 occurrence is at a degree above average in the number of searches and below average in the number of hyperlink navigations for 1 Web page; and 1 occurrence at a degree below average in the number of searches, and below average in the number of hyperlink navigations for 4 Web pages.

Although our research focuses on two dimensional fuzzy bags since it is a direct application of the 2 dimensional nature of our query (see abstract), more research is needed in the area of n dimensional fuzzy bags. The authors believe that this becomes critical as search engines are used to query complex multimedia scenes and images. More especially, the association of n fuzzy bags in section 4 needs to be revisited and applied to n dimensional queries.

This study further introduces a similarity measure between successful and unsuccessful users. The similarity measure shows that the successful users were more similar to unsuccessful users on the one-dimensional query than the 2- dimensional query. User's searching behavior on the WWW represents a major challenge to many intelligent approaches developed already or to be developed to meet such a challenge. The authors believes that more intelligent and soft approaches are needed to mine the behavior of users searching the WWW. The authors of this study encourage other user's Web behavior experts to use this study as a model to build on in their actual data and observed attributes. Research along these lines is essential to ensuring that this tool is properly integrated with other emerging technologies to provide World

Wide Web user's behavior experts' ways to better interpret the behavior of users whether children or adults while searching the Web. The complexity of the WWW makes intelligent approaches and techniques to explore the behavior of users in such a space an important task. Although other intelligent approaches exist and are being applied to soft domains, Rough set theory [98] could be a good candidate to meet such a challenge for studying users searching the WWW. As Rough set theory is being applied successfully in a variety of many real-world problems ranging from medicine to market analysis, applying it to the data mined from user's behavior while searching the WWW will help better their behavior and their strategy in forming concepts and refining their queries to answer questions on the WWW.

6.5 A Rough Set Theory to Interpret User's Web Behavior

6.5.1 General Introduction to Rough Set Theory and Decision Analysis

The Rough set approach to data analysis and modeling ([106–109]) has the following advantages:

- It is based on the original data and does not need any external information (probability or grade of membership);
- It is a suitable tool for analyzing quantitative and qualitative attributes;
- It provides efficient algorithms for finding hidden patterns in data;
- It finds minimal sets of data (data reduction);
- It evaluates the significance of data.

We show that the Rough set theory is a useful tool for analysis of decision situations, in particular multi-criteria sorting problems. It deals with vagueness in representation of a decision situation, caused by granularity of the representation. The Rough set approach produces a set of decision rules involving a minimum number of most important criteria. It does not correct vagueness manifested in the representation; instead, produced rules are categorized into deterministic and non deterministic. The set of decision rules explains a decision policy and may be used for decision support. Mathematical decision analysis intends to bring to light those elements of a decision situation that are not evident for actors and may influence their attitude towards the situation. More precisely, the elements revealed by the mathematical decision analysis either explain the situation or prescribe, or simply suggest, some behavior in order to increase the coherence between evolution of the decision process on the one hand and the goals and value system of the actors on the other. A formal framework for discovering facts from representation of a decision situation called Rough set theory. Rough set theory assumes the representation in a decision table in which there is a special case of an information system. Rows of this table correspond to objects (actions, alternatives, candidates, patients, ···) and columns correspond to attributes. For each pair (object, attribute) there is a known value called a descriptor. Each row of the table contains descriptors representing information about the corresponding object of a given decision situation. In general, the set of attributes is partitioned into two subsets: condition attributes (criteria, tests, symptoms, ···) and decision attributes (decisions, classifications, taxonomies, ···). As in decision problems the concept of criterion is often used instead of condition attribute; it should be noticed that the latter is more general than the former because the domain (scale) of a criterion has to be ordered according to decreasing or increasing preference while the domain of a condition attribute need not be ordered. Similarly, the domain of a decision attribute may be ordered or not. In the case of a multi-criteria sorting problem, which consists in assignment of each object to an appropriate predefined category (for instance, acceptance, rejection or request for additional information), Rough set analysis involves evaluation of the importance of particular criteria:

- construction of minimal subsets of independent criteria
- having the same discernment ability as the whole set;
- non-empty intersection of those minimal subsets to give a core of criteria which cannot be eliminated without it;
- disturbing the ability of approximating the decision;
- elimination of redundant criteria from the decision table;
- the generation of sorting rules (deterministic or not) from the reduced decision table, which explain a decision;
- Development of a strategy which may be used for sorting new objects.

6.5.2 Rough Set Modeling of User Web Behavior

The concept of Rough set theory is based on the assumption that every object of the universe of discourse is associated with some information. Objects characterized by the same information are indiscernible in view of their available information. The indiscernibility relation generated in this way is the mathematical basis of Rough set theory. The concepts of Rough set and fuzzy set are different since they refer to various aspects of non-precision. Rough set analysis can be used in a wide variety of disciplines; wherever large amounts of data are being produced, Rough sets can be useful. Some important application areas are medical diagnosis, pharmacology, stock market prediction and financial data analysis, banking, market research, information storage and retrieval systems, pattern recognition (including speech and handwriting recognition), control system design, image processing and many others. Next, we show some basic concepts of Rough set theory. The application discussed in this paper is based on a fact based query "Limbic functions of the

Table 6.18. Summary of All Users

Users	W	H	SE	E
1	L	M	M	F
2	S	S	S	SU
3	M	L	L	SU
4	L	L	VL	F
5	L	L	L	F
6	M	M	L	SU
7	S	M	S	F
8	L	L	M	F
9	M	S	L	F
10	M	S	M	SU
11	S	M	M	SU
12	M	S	M	SU
13	M	M	L	SU
14	VL	M	VL	SU
15	S	S	S	SU
16	S	M	S	SU
17	S	S	S	SU
18	S	S	S	SU
19	S	S	S	F
20	S	S	M	F

brain"(Meghabghab [97]). The results of the query is summarized in Table 6.18 but with the following coded values: (S=1,2;M=3,4;L=5,7;VL=8,9,10).

The notion of a User Modeling System presented here is borrowed from ([106]). The formal definition of a User Modeling System (UMS) is represented by S=(U, Ω, V, f) where: U is a non-empty, finite set of users called the universe;Ω is a non-empty, finite set of attributes: $C \cup D$, in which C is a finite set of condition attributes and D is a finite set of decision attributes; $V = \bigcup V_q$ is a non empty set of values of attributes, and V_q is the domain of q ($\forall q \in \Omega$); f is a User Modeling Function :

$$f : U \times \Omega \to V \qquad (6.53)$$

such that:$\exists$ f(q,p) $\in V_p$ $\forall$ p $\in$ U and q $\in \Omega$

$$f_q : \Omega \to V \qquad (6.54)$$

such that: $\exists f_q(p) = f(q, p)$ $\forall$ p $\in$ U and q $\in \Omega$ is the user knowledge of U in S. This modeling system can be represented as a table in which columns represent attributes and rows are users and an entry in the qth row and p^{th} column has the value f(q,p). Each row represents the user's attributes in answering the question. In what follows is the outline of the rest of the paper. We will study 3 cases of the fact based query "Limbic functions of the brain" (Meghabghab [97]):

- The case of the 3 condition variables or attribute variables: Ω = {Searches, hyperlinks, Web Pages }= {SE,H,W} (Meghabghab [98]). We will reduce the number of attributes and generate the rules that explain the behavior of the users using these 3 variables.
- The case of the 3 condition variables or attribute variables: Ω = {Web Pages,Time,Searches}= {W,T,SE}. We will reduce the number of attributes and generate the rules that explain the behavior of the users using these 3 variables.
- The case of the 4 condition variables or attribute variables: Ω = {Web Pages,Hyperlinks,Time,Searches} = {W,H,T,SE}. We will reduce the number of attributes and generate the rules that explain the behavior the users using these 4 variables.
- Final Comparison between Rough Set and ID_3 on all three cases.

Rough Set applied to Case 1: Searches, hyperlinks, and Web Pages([98]
Consider the following example:
U={1,2,3,4,5,6,7,8,9,10,11,12,13,14,15,16,17,18,19,20}= set of 20 users who searched the query, Ω ={Searches, hyperlinks, Web Pages}={SE,H,W}= set of 3 attributes. Since some users did search more than others, browsed more than others, scrolled down Web pages more than others, a simple transformation of table 16 yields a table up to 3 different attributes with a set of values ranging form: small(S), medium(M), large(L), and very large(VL).

$\Omega = \{SE, H, W\}$,
V_{SE}={M,S,L,VL,L,L,S,M,L,M,M,M,L,VL,S,S,S,S,S,M},
V_H={M,S,L,L,L,M,M,M,L,S,S,M,S,M,M,S,M,S,S,S,S},
V_W={L,S,M,L,L,M,S,L,M,M,S,M,M,VL,S,S,S,S,S,S}.

The modeling system will now be extended by adding a new column E representing the expert's evaluation of the user's knowledge whether the user's succeeded in finding the answer or failed to find the answer. In a new UMS, S is represented by S=(U, Ω, V, f), $f_q(p)$ where q $\in \Omega$ and p $\in$ PU={P-E} is the user's knowledge about the query, and $f_q(p)$ where q $\in \Omega$ and p = E is the expert's evaluation of the query for a given student. E is the decision attribute. Consider the above example but this time: $\Omega = \Omega_u \cup \Omega_e$ = {SE,H,W} $\cup$ E, where E= SU(success) or F(failure);

V_{SE}={M,S,L,VL,L,L,S,M,L,M,M,M,L,VL,S,S,S,S,S,M},
V_H={M,S,L,L,L,M,M,M,L,S,S,M,S,M,M,S,M,S,S,S,S},
V_W={L,S,M,L,L,M,S,L,M,M,S,M,M,VL,S,S,S,S,S,S}
V_E={F,SU,SU,F,F,SU,F,F,F,SU,SU,SU,SU,SU,SU,SU,SU,SU,F,F}.

Lower and Upper Approximations
In Rough set theory the approximations of a set are introduced to deal with indiscernibility. If S= (U, Ω, V, f) is a decision table, and X $\in$ U, then the I_* lower and I^* upper approximations of X are defined, respectively, as follows:

$$I_* = \{x \in U, I(x) \subseteq X\} \tag{6.55}$$

$$I^* = \{x \in U, I(x) \cap X \neq \emptyset\} \tag{6.56}$$

where I(x) denotes the set of all objects indiscernible with x, *i.e.*, equivalence class determined by x. The boundary region of X is the set $BN_I(X)$:

$$BN_I(X) = I_*(X) - I^*(X) \tag{6.57}$$

Here are different values for $BN_I(X)$:

- If the boundary region of X is the empty set, i.e., $BN_I(X) = \emptyset$, then the set X will be called crisp with respect to I;
- in the opposite case, i.e., if $BN_I(X) \neq \emptyset$, the set X will be referred to as Rough with respect to I. Vagueness can be characterized numerically by defining the following coefficient:

$$I(X) = |I_*(X)|/|I^*(X)| \tag{6.58}$$

where $|X|$ denotes the cardinality of the set X.

Obviously $0 < \alpha_I(X) \leq 1$. If $\alpha_I(X)=1$ the set X is crisp with respect to I; otherwise if $\alpha_I(X) < 1$, the set X is Rough with respect to I. Thus the coefficient I(X) can be understood as the accuracy of the concept X. Assuming that I_1 and I_2 are equivalence relations in U, the positive region concept $POS_{I_1}(I_2)$ is defined as:

$$POS_{I1}(I_2) = \bigcup_{x \in I_2} I_1(X) \tag{6.59}$$

A positive region contains all patterns in U that can be classified in the attribute set I_2 using the information in attribute set I_2. The degree of dependency of $\Omega = \{W,H,SE\}$ with respect to D=$\{$ S $\}$ is:

$$\gamma(\Omega, D) = |(POS_\Omega(D))|/|(U)| \tag{6.60}$$

Or also:

$$\gamma(\Omega, D) = |\cup_{x \in D} \Omega_*(X)|/|(U)| \tag{6.61}$$

The degree of dependency provides a measure of reducing the number of attributes in Ω without compromising the decision made in the decision table.

Indiscernibility

Let $\Gamma =$(S,U) be an information system, then with any B $\subseteq$ S, there is an associated equivalence relation $Ind_\Gamma(B)$:

$$Ind_\Gamma(B) = \{(x, x') \in U^2 | \forall a \in B \ a(x) = a(x')\} \tag{6.62}$$

$Ind_\Gamma(B)$ is called the B-Indiscernibility relation.

If (x,x') $\in Ind_\Gamma(B)$ then objects x and x' are indiscernible from each other by attributes from B. Of course B can be one attribute or all the attributes of S. The equivalence classes of the B-Indiscernibility relation are denoted $[x]_B$. The subscript in the Indiscernibility relation is omitted since we know which system we are talking about. Let us illustrate how a decision table such as table 16 defines an indiscernibility relation. By applying 6.61 to table 16 yields the following:

Ind({W})={{2,7,11,15,16,17,18,19,20}$_S$,{3,6,9,10,12,13}$_M$,{1,4,5,8}$_L$, {14}$_{VL}$}

Ind({H})={{2,9,10,12,15,17,18,19,20}$_S$,{1,6,7,11,13,14,16}$_M$,{3,4,5,8}$_L$}

Ind({SE})={{2,7,15,16,17,18,19}$_S$,{1,8,10,11,12,20}$_M$,{3,5,6,9,13}$_L$, {4,14}$_{VL}$}

If B were to be have more than 1 attribute, we can calculate the Ind of 2 attributes B_1 and B_2 as the intersection of the Ind of each one of the attribute B_1 and B_2 as follows:

$$Ind_\Gamma(B) = Ind(B_1, B_2) \tag{6.63}$$

$$Ind_\Gamma(B_1 \cap B_2) = \{X \cap Y : \forall X \in Ind(B_1) and Y \in Ind(B_2), X \cap Y \neq \emptyset$$

By applying 6.62 to the already calculated Ind of the individual attributes W, H, and SE, we can calculate the followings:

Ind({W,H})={{2,15,17,18,19,20},{7,11,16},{9,10,12},{6,13},{3}, {4,5,8}, {1},{14}}

Ind({W,SE})={{2,7,15,16,17,18,19},{11,20},{10,12},{3,6,9,13}, {1,8},{5}, {4},{14}}

Ind({H,SE})={{2,15,17,18,19},{10,12,20},{7,16},{1,11},{6,13}, {8},{9}, {3,5},{4},{14}}

Ind({W,H,SE})={{2,15,17,18,19},{7,16},{10,12},{6,13},{3},{20},{1}, {8},{9},{5},{4},{11},{14}}

A vague concept has boundary-line cases, i.e., elements of the universe which cannot be - with certainty- classified as elements of the concept. Here uncertainty is related to the question of membership of elements to a set. Therefore in order to discuss the problem of uncertainty from the Rough set perspective we have to define the membership function related to the Rough set concept (the Rough membership function). The Rough membership function can be defined employing the indiscernibility relation I as:

$$\mu_X^I(x) = |X \cap I(x)|/|I(x)| \tag{6.64}$$

Obviously, $0 < \mu_X^I(x) \leq 1$. The Rough membership function can be used to define the approximations and the boundary regions of a set, as shown below:

$$I_*(X) = x \in U : \mu_X^I(x) = 1 \tag{6.65}$$

$$I^*(X) = x \in U : \mu_X^I(x) > 1 \tag{6.66}$$

$$BN_I(X) = x \in U : \mu_X^I(x) < 1 \tag{6.67}$$

Once can see from the above definitions that there exists a strict connection between vagueness and uncertainty in the Rough set theory. As we mentioned above, vagueness is related to sets, while uncertainty is related to elements of sets. Thus approximations are necessary when speaking about vague concepts, whereas Rough membership is needed when uncertain data are considered.

Application of the above definitions of Rough set theory to Table 6.18

In table 6.18, users {2,7,15,16,17,18,19} are indiscernible according to the attribute SE=S, users {1,4,5} are indiscernible for the attribute W=L. For example the attribute W generates 4 sets of users: $\{2,7,11,15,16,17,18,19,20\}_S$, $\{3,6,9,10,12,13\}_M$, $\{1,4,5,8\}_L$, and $\{14\}_{VL}$. Because users {2,15,17,18} were SU and user {19} failed, and are indiscernible to attributes W=S, H=S, and SE=S, then the decision variable for SU or F cannot be characterized by W=S, H=S, and SE=S. Hence users {2,15,17,18} and {19} are boundary-line cases. Because user {16} was successful (SU) and user {7} has failed (F), and they are indiscernible to attributes W=S, H=M, and SE=S, then the decision variable for SU or F cannot be characterized by W=S, H=M, and SE=S. Hence users {16} and {7} are boundary-line cases. The remaining users: {3,6,10, 11,12,13,14} have characteristics that enable us to classify them as being SU, while users {1,4,5,8,9,20} display characteristics that enable us to classify them as F, and users {2,7,15,16,17,18,19} cannot be excluded from being SU or F. Thus the lower approximation of the set of being SU by applying 6.55 is: $I_*(SU)=\{3,6,10,11,12,13,14\}$ and the upper approximation of being SU by applying 6.56 is: $I^*(SU)=\{2,7,15,16,17,18,19,3,6,10,11,12,13,14\}$.

Similarly in the concept of F, its lower approximation by applying 6.55 is:

$I_*(F)=\{1,4,5,8,9,20\}$ and its upper approximation by applying 6.56 is:

$I^*(F)=\{1,4,5,8,9,20,3,6,10,11,12,13,14\}$.

The boundary region of the set SU or F by applying (6.57) is:

$BN_I(SU)=BN_I(F)=\{2,7,15,16,17,18,19\}$.

The positive region of being SU and F by applying (6.59) is: $I_*(F) \cup I_*(SU) = \{1,3,4,5,6,8,9,10,11,12,13,14,20\} \neq U\ I^*(F) \cup I^*(SU)$ is called the I-positive region of S and is denoted by $POS_I(SE)$.

Figure 6.27 shows an equivalence relation of both successful and non successful users using Rough Set Theory. By applying (6.58) to SU:
$\alpha(SU)=|\{3,6,10,11,12,13,14\}|/|\{2,7,15,16,17,18,19,3,6,10,11,12,13,14\}| = 7/14 = .5$.

We also compute the membership value of each user to the concept of "SU" or "F". By applying (6.63) to SU we have:

$\mu(SU)(1) = |\{2,3,6,10,11,12,13,14,15,16,17,18\} \cap \{1\}|/|\{1\}| = 0$

$\mu(SU)(2) = |\{2,3,6,10,11,12,13,14,15,16,17,18\} \cap \{2,7,15,16,17,18,19\}|$
$/|\{2,7,15,16,17,18,19\}| = .71$

$\mu(SU)(3) = |\{2,3,6,10,11,12,13,14,15,16,17,18\} \cap \{3\}|/|\{3\}| = |\{3\}|$
$/\{3\}| = 1$

$\mu(SU)(4) = |\{2,3,6,10,11,12,13,14,15,16,17,18\} \cap \{4\}|/|\{4\}| = 0$

$\mu(SU)(5) = |\{2,3,6,10,11,12,13,14,15,16,17,18\} \cap \{5\}|/|\{5\}| = 0$

$\mu(SU)(6) = |\{2,3,6,10,11,12,13,14,15,16,17,18\} \cap \{6\}|/|\{6\}| = |\{6\}|/|\{6\}|$
$= 1$

$\mu(SU)(7) = |\{2,3,6,10,11,12,13,14,15,16,17,18\} \cap \{2,7,15,16,17,18,19\}|$
$/|\{2,7,15,16,17,18,19\}| = .71$

$\mu(SU)(8) = |\{2,3,6,10,11,12,13,14,15,16,17,18\} \cap \{8\}| = 0$

$\mu(SU)(9) = |\{2,3,6,10,11,12,13,14,15,16,17,18\} \cap \{9\}|/|\{9\}| = 0$

$\mu(SU)(10) = |\{2,3,6,10,11,12,13,14,15,16,17,18\} \cap \{10\}|/|\{10\}| =$
$|\{10\}|/|\{10\}| = 1$

$\mu(SU)(11) = |\{2,3,6,10,11,12,13,14,15,16,17,18\} \cap \{11\}|/|\{11\}| =$
$|\{11\}|/|\{11\}| = 1$

$\mu(SU)(12) = |\{2,3,6,10,11,12,13,14,15,16,17,18\} \cap \{12\}|/|\{12\}| = 1$

$\mu(SU)(13) = |\{2,3,6,10,11,12,13,14,15,16,17,18\} \cap \{13\}|/|\{13\}| = 1$

$\mu(SU)(14) = |\{2,3,6,10,11,12,13,14,15,16,17,18\} \cap \{14\}|/|\{14\}| = 1$

$\mu(SU)(15) = |\{2,3,6,10,11,12,13,14,15,16,17,18\} \cap \{2,7,15,16,17,$
$18,19\}|$
$/|\{2,7,15,16,17,18,19\}| = 5/7$

$\mu(SU)(16) = |\{2,3,6,10,11,12,13,14,15,16,17,18\} \cap \{2,7,15,16,17,$
$18,19\}|$
$/|\{2,7,15,16,17,18,19\}| = 5/7$

$\mu(SU)(17) = |\{2,3,6,10,11,12,13,14,15,16,17,18\} \cap \{2,7,15,16,17,$
$18,19\}|$
$/|\{2,7,15,16,17,18,19\}| = 5/7$

$\mu(SU)(18) = |\{2,3,6,10,11,12,13,14,15,16,17,18\} \cap \{2,7,15,16,17,$
$18,19\}|$
$/|\{2,7,15,16,17,18,19\}| = 5/7$

$\mu(SU)(19) = |\{2,3,6,10,11,12,13,14,15,16,17,18\} \cap \{2,7,15,16,17,$
$18,19\}|$
$/|\{2,7,15,16,17,18,19\}| = 5/7$

$\mu(SU)(20) = |\{2,3,6,10,11,12,13,14,15,16,17,18\} \cap \{20\}| = 0$

$\mu(SU) = \{0/1, .71/2, 1/3, 0/4, 0/5, 1/6, .71/7, 0/8, 0/9, 1/10, 1/11, 1/12, 1/13,$
$1/14, .71/15, .71/16, .71/17, .71/18, .71/19, 0/20\}$

By applying (58) to F:

$\alpha(F) = |\{1,4,5,8,9,20\}|/|\{2,7,15,16,17,18,19,1,4,5,8,9,20\}| = 6/13 = .45$

By applying (63) we have F as follows:

$\mu(F)(1) = |\{1,4,7,8,9,19,20\} \cap \{1\}|/|\{1\}| = 1$

$\mu(F)(2) = |\{1,4,7,8,9,19,20\} \cap \{2,7,15,16,17,18,19\}|/|\{2,7,15,16,17, 18,19\}| = 2/7$

$\mu(F)(3) = |\{1,4,7,8,9,19,20\} \cap \{3\}|/|\{3\}| = 0$

$\mu(F)(4) = |\{1,4,7,8,9,19,20\} \cap \{4\}|/|\{4\}| = 1$

$\mu(F)(5) = |\{1,4,7,8,9,19,20\} \cap \{5\}|/|\{5\}| = 1$

$\mu(F)(6) = |\{1,4,7,8,9,19,20\} \cap \{6\}|/|\{6\}| = 0$

$\mu(F)(7) = |\{1,4,7,8,9,19,20\} \cap \{2,7,15,16,17,18,19\}|/|\{2,7,15,16,17, 18,19\}| = 2/7$

$\mu(F)(8) = |\{1,4,7,8,9,19,20\} \cap \{8\}|/|\{8\}| = 1$

$\mu(F)(9) = |\{1,4,7,8,9,19,20\} \cap \{9\}|/|\{9\}| = 1$

$\mu(F)(10) = |\{1,4,7,8,9,19,20\} \cap \{10\}|/|\{10\}| = 0$

$\mu(F)(11) = |\{1,4,7,8,9,19,20\} \cap \{11\}|/|\{11\}| = 0$

$\mu(F)(12) = |\{1,4,7,8,9,19,20\} \cap \{12\}|/|\{12\}| = 0$

$\mu(F)(13) = |\{1,4,7,8,9,19,20\} \cap \{13\}|/|\{13\}| = 0$

$\mu(F)(14) = |\{1,4,7,8,9,19,20\} \cap \{14\}|/|\{14\}| = 0$

$\mu(F)(15) = |\{1,4,7,8,9,19,20\} \cap \{2,7,15,16,17,18,19\}|/|\{2,7,15,16,17, 18,19\}| = 2/7$

$\mu(F)(16) = |\{1,4,7,8,9,19,20\} \cap \{2,7,15,16,17,18,19\}|/|\{2,7,15,16,17, 18,19\}| = 2/7$

$\mu(F)(17) = |\{1,4,7,8,9,19,20\} \cap \{2,7,15,16,17,18,19\}|/|\{2,7,15,16,17, 18,19\}| = 2/7$

$\mu(F)(18) = |\{1,4,7,8,9,19,20\} \cap \{2,7,15,16,17,18,19\}|/|\{2,7,15,16,17, 18,19\}| = 2/7$

$\mu(F)(19) = |\{1,4,7,8,9,19,20\} \cap \{2,7,15,16,17,18,19\}|/|\{2,7,15,16,17, 18,19\}| = 2/7$

$\mu(F)(20) = |\{1,4,7,8,9,19,20\} \cap \{20\}|/|\{20\}| = 1$

$\mu(F) = \{1/1, .29/2, 0/3, 1/4, 1/5, 0/6, .29/7, 1/8, 1/9, 0/10, 0/11, 0/12, 0/13, 0/14, .29/15, .29/16, .29/17, .29/18, .29/19, 1/20\}$

Figure 6.26 shows both Rough membership function of both success and failure, F and SU.

Decision Rules according to Rough Set Theory

Usually we need many classification patterns of objects. For example users can be classified according to Web pages, number of searches, ···. Hence we can assume that we have not one, but a family of indiscernibility relations

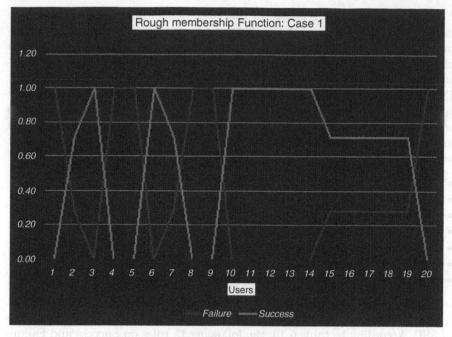

Fig. 6.26. Case 1

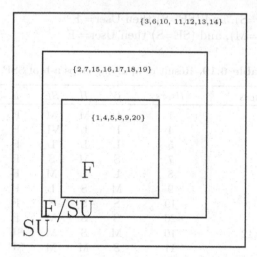

Fig. 6.27. Approximating SU and F using W, H, and SE

$\{I_1, I_2, I_3, \cdots, I_n\}$ over the universe U. The set theoretical intersection of equivalence relations $\{I_1, I_2, I_3, \cdots, I_n\}$ is denoted by:

$$\bigcap I = \bigcap_{i=1}^{n}(I_i) \qquad (6.68)$$

I is also an equivalence relation. In this case, elementary sets are equivalence classes of the equivalence relation I. Because elementary sets uniquely determine our knowledge about the universe, the question arises whether some classification patterns can be removed without changing the family of elementary sets- or in other words, preserving the indiscernibility. Minimal subset I' of I such that will be called a reductof I. Of course we can have many reducts. Finding reducts is not a very simple task and there are methods to solve this problem. The algorithm we use has been proposed by (Slowinski and Stefanowski [120]), and it is summarized by the following procedure that we name SSP:

- Transform continuous values in ranges;
- Eliminate identical attributes;
- Eliminate identical examples;
- Eliminate dispensable attributes;
- Calculate the core of the decision table;
- Determine the reductset;
- Extract the final set of rules.

Applying the SSP procedure from steps a-b results in:

The number of users is reduced from 20 to 15 users because of steps a-b of SSP. According to table 6.19, the following 15 rules on Success and Failure can be made by just applying steps a-b:

1. If (W=S), (H=S), and (SE=S) then User= F
2. If (W=S), (H=M), and (SE=S) then User= F

Table 6.19. Result of applying steps a-b of SSP

Users	Users	W	H	SE	E
1	1	L	M	M	F
4	4	L	L	VL	F
5	5	L	L	L	F
7	7	S	M	S	F
8	8	L	L	M	F
9	9	M	S	L	F
19	19	S	S	S	F
20	20	S	S	M	F
10,12	10	M	S	M	SU
11	11	S	M	M	SU
12	12	M	S	M	SU
6,13	6	M	M	L	SU
14	14	VL	M	VL	SU
2,15,17,18	2,15,17,18	S	S	S	SU
16	16	S	M	S	SU
17	17	S	S	S	SU
18	18	S	S	S	SU

3. If (W=S), (H=S), and (SE=M) then User= F
4. If (W=M), (H=S), and (SE=L) then User= F
5. If (W=L), (H=M), and (SE=M) then User= F
6. If (W=L), (H=L), and (SE=M) then User= F
7. If (W=L), (H=L), and (SE=L) then User= F
8. If (W=L), (H=L), and (SE=VL) then User= F
9. If (W=S), (H=S), and (SE=S) then User= SU
10. If (W=S), (H=S), and (SE=M) then User= SU
11. If (W=S), (H=M), and (SE=M) then User= SU
12. If (W=M), (H=S), and (SE=M) then User= SU
13. If (W=M), (H=M), and (SE=L) then User= SU
14. If (W=M), (H=L), and (SE=L) then User= SU
15. If (W=VL), (H=M), and (SE=VL) then User= SU

Since users {2,7,15,16,17,18} and {19} are indiscernible they can be made out into a contradictory set of rules and reduce the number of users from 15 to 11 users. The 11 deterministic rules extracted from table 6.20 as it applies to success and failure are:

- If (W=S), (H=M), and (SE=S) then User= F
- If (W=M), (H=S), and (SE=L) then User= F
- If (W=L), (H=M), and (SE=M) then User= F
- If (W=L), (H=L), and (SE=M) then User= F
- If (W=L), (H=L), and (SE=L) then User= F
- If (W=L), (H=L), and (SE=VL) then User= F
- If (W=S), (H=S), and (SE=M) then User= SU
- If (W=S), (H=M), and (SE=M) then User= SU
- If (W=M), (H=M), and (SE=L) then User= SU
- If (W=M), (H=L), and (SE=L) then User= SU

Table 6.20. Eliminating all contradictory users

Users	Users	W	H	SE	E
20	20	S	S	M	F
1	1	L	M	M	F
8	8	L	L	M	F
9	9	M	S	L	F
5	5	L	L	L	F
4	4	L	L	VL	F
10,12	10	M	S	M	SU
11	11	S	M	M	SU
6,13	6	M	M	L	SU
3	3	M	L	L	SU
14	14	VL	M	VL	SU

Table 6.21. 9 users are left after applying steps a-d of SSP

Users	Users	W	H	SE	E
20	20	S	S	M	F
1	1	L	M	M	F
8	8	L	L	M	F
9	9	M	S	L	F
5	5	L	L	L	F
4	4	L	L	VL	F
11	11	S	M	M	SU
3	3	M	L	L	SU
14	14	VL	M	VL	SU

- If (W=VL), (H=M), and (SE=VL) then User= SU

By eliminating all identical examples we have nine users left out of the original 20 users (see table 6.18)

The 9 deterministicrules extracted from table 6.21 as it applies to success and failure are:

- If (W=S), (H=M), and (SE=S) then User= F
- If (W=M), (H=S), and (SE=L) then User= F
- If (W=L), (H=M), and (SE=M) then User= F
- If (W=L), (H=L), and (SE=M) then User= F
- If (W=L), (H=L), and (SE=L) then User= F
- If (W=L), (H=L), and (SE=VL) then User= F
- If (W=S), (H=M), and (SE=M) then User= SU
- If (W=M), (H=L), and (SE=L) then User= SU
- If (W=VL), (H=M), and (SE=VL) then User= SU

Discernibiliy Matrix and Discernibility Function

Let A be an information system with n objects. The Discernibility Matrix of A is a n × n symmetric matrix with entries c_{ij} as given below. Each entry thus consists of the set of attributes upon which objects x_i and x_j differ:

$$c_{ij} = \{a \in A | a(x_i) \neq a(x_j)\} \forall i, j = 1, 2, \cdots \qquad (6.69)$$

A Discernibiliy function f_A in an information system A, is a boolean function of m variables $a_1^*, a_2^*, a_3^*, \cdots, a_m^*$ corresponding to the attributes $a_1^*, a_2^*, a_3^*, \cdots, a_m^*$ defined as below, where $c_{ij}^* = \{a^* | a \in c_{ij}\}$.

$$f_A(a_1^*, a_2^*, a_3^*, \cdots, a_m^*) = \bigcap \{\bigcup c_{ij}^* | 1 < j < i < n; c_{ij} \neq 0\} \qquad (6.70)$$

Example. Applying (6.69) to Table 6.22 yields the following Discernibility Function:

Table 6.22. Reordered decision table of table 6.21

Users	W	H	SE	E
3	M	L	L	SU
11	S	M	M	SU
14	VL	M	VL	SU
1	L	M	M	F
4	L	L	VL	F
5	L	L	L	F
8	L	L	M	F
9	M	S	L	F
20	S	S	M	F

$$f_A = \{(W \cup H \cup SE)(W \cup SE)(W)(W \cup SE)(H)(W \cup H \cup SE)$$
$$(W)(W \cup H \cup SE)(W \cup H \cup SE)(W \cup H)(W \cup H \cup SE)(H)$$
$$(W \cup SE)(W \cup H)(H \cup SE)(W \cup H \cup SE)(W \cup H \cup SE)$$
$$(W \cup H \cup SE)\}$$
$$f_A = \{(HW) \cap (HW) \cap ((W \cup H) \cap (W \cup SE) \cap (H \cup SE))\}$$
$$f_A = \{(HW) \cap ((W \cup H) \cap (W \cup SE) \cap (H \cup SE))\}$$
$$f_A = HW$$

Where HW stands for: $(H \cap W)$.

Thus the 3 attributes can be reduced without compromising the discernibility of the decision table 6.21. 2 attributes H and W are just sufficient to make the decision as whether a user succeeded or failed. The attribute SE is become superfluous. Applying (6.68) yields a symmetric matrix (9×9) which is half filled:

	3	11	14	1	4	5	8	9	13
3	$\emptyset$								
11	$\emptyset$	$\emptyset$							
14	$\emptyset$	$\emptyset$	$\emptyset$						
1	W,H,SE	W	W,SE	$\emptyset$					
4	W,SE	W,H,SE	W	$\emptyset$	$\emptyset$				
5	W	W,H,SE	H,SE	$\emptyset$	$\emptyset$	$\emptyset$			
8	W,SE	W	W,H,SE	$\emptyset$	$\emptyset$	$\emptyset$	$\emptyset$		
9	H	W,H,SE	W,H,SE	$\emptyset$	$\emptyset$	$\emptyset$	$\emptyset$	$\emptyset$	
13	W,H,SE	H	W,H,SE	$\emptyset$	$\emptyset$	$\emptyset$	$\emptyset$	$\emptyset$	$\emptyset$

Table 6.23 shows the attributes W and H knowing that 2 attributes are enough. The result of applying of steps a through b to table 6.23 is shown in table 6.24.

The result of applying of steps e through f and knowing that 2 attributes are more than enough is displayed in tables 6.25 and 6.26. (X stands for any value).

Table 6.23. Table 6.22 with only 2 attributes W and H

Users	W	H	E
3	M	L	SU
11	S	M	SU
14	VL	M	SU
1	L	M	F
4	L	L	F
5	L	L	F
8	L	L	F
9	M	S	F
20	S	S	F

Table 6.24. Applying steps a-b to table 6.23

Users	W	H	E
3	M	L	SU
11	S	M	SU
14	VL	M	SU
1	L	M	F
9	M	S	F
20	S	S	F

Table 6.25. Core of the set of final data (Step e- of SSP)

Users	W	H	E
20	X	S	F
9	X	S	F
1	L	X	F
3	X	L	SU
11	X	M	SU
14	VL	X	SU

Table 6.26. Set of reductset (Step f- of SSP)

Users	W	H	E
20,9	X	S	F
1	L	X	F
3	X	L	SU
11	X	M	SU
14	VL	X	SU

Deterministic Rules extracted

1. 3 Rules on Success:

   (a) If (W=M and H=L) then User=SU
   (b) If (W=VL) then User= SU
   (c) (W=S and H=M) then User= SU

2. 3 Rules on Failure:

   (a) If (W=M and H=S) then User= F
   (b) If (W=L) then User= F
   (c) If (W=S and H=S) then User= F

These 6 rules say that only 2 parameters are needed to classify the 6 deterministic examples or users out of the original 20, and they are W,H. It seems that these deterministic rules are independent of the number of searches or the attribute SE.

Reducts, reduct Sets, and Minimum reductSet

Computing equivalence classes is straightforward. Finding e.g. minmimal reducts (i.e. reductwith minimum cardinality among all reducts) is NP -hard (Skowron and Rauszer [119]). It means computing reducts is a no trivial task that cannot be solved by a simple increase of computational resources. In fact, it is one of the bottlenecks of Rough set methodology. A reduct of Γ is a minimal set of attributes B $\subset$ A such that Ind(B)=Ind(A). In other words a reduct is a minimal set of attributes that preserves the partitioning of the universe, and hence to perform the classification as the whole set of attributes set A does. A decision table may have more than one reduct. Any reduct can be used to replace the original table. Thus any reduct is sufficient. A question arises which is one is the best since there could be more than one. The selection is based on an "optimality" criterion associated with the attributes. Thus a minimum cost criterion function can be calculated. Given a classification task mapping a set of attributes Ω to a decision d of D, a reductset RS with respect to the power set of Ω, e.g., Power(Ω) or 2^{Ω} such that and using equation (6.60) or (6.61):

$$RS = \{A \in 2^{\Omega} : \gamma(A, D) = \gamma(\Omega, D)\}$$

Where $\gamma(\Omega, D)$ is defined in equation (6.60) or (6.61). In our example $\Omega = \{W,H,SE\}$ and D=\{S\}, and thus the power set of Ω is equal to: 2^{Ω} =\{\emptyset, \{W\}, \{H\}, \{SE\}, \{WH\}, \{WSE\}, \{HSE\}, \{WHSE\}\}.

K is said to be a minmimal RS or MRS iff it is a reductset RS with the minimum number of elements in it such that :

$$|K| \le |R|, where R \in 2^{\Omega}$$

Applying the concept of minimal reduct set to find deterministic rules Thus the lower approximation of the set of being SU by applying (6.55) to table 6.21 is: $I_*(SU)=\{3,11,14\}$ and the upper approximation of being SU by applying

(6.56) is: $I^*(SU)=\{3,11,14\}$. Similarly in the concept of F, its lower approximation by applying (6.55) is: $I_*(F)=\{1,4,5,8,9,20\}$ and its upper approximation
by applying (6.56) is: $I^*(F)=\{1,4,5,8,9,20\}$. Table 6.21 has no boundary regions. By applying (6.62) and (6.63) to table 6.20, as we did above, yields the
following:

$$\mathrm{Ind}(\{W\}) = \{\{11,20\}_S, \{3,9\}_M, \{1,4,5,8\}_L, \{14\}_{VL}\}$$
$$\mathrm{Ind}(\{H\}) = \{\{9,20\}_S, \{1,11,14\}_M, \{3,4,5,8\}_L\}$$
$$\mathrm{Ind}(\{SE\}) = \{\{1,8,11,20\}_M, \{3,5,9\}_L, \{4,14\}_{VL}\}$$
$$\mathrm{Ind}(\{W,H\}) = \{\{20\}, \{11\}, \{9\}, \{3\}, \{4,5,8\}, \{1\}, \{14\}\}$$
$$\mathrm{Ind}(\{W,SE\}) = \{\{11,20\}, \{3,9\}, \{1,8\}, \{5\}, \{4\}, \{14\}\}$$
$$\mathrm{Ind}(\{H,SE\}) = \{\{20\}, \{1,11\}, \{8\}, \{9\}, \{3,5\}, \{4\}, \{14\}\}$$
$$\mathrm{Ind}(\{W,H,SE\}) = \{\{3\}, \{20\}, \{1\}, \{8\}, \{9\}, \{5\}, \{4\}, \{11\}, \{14\}\}$$

Applying (6.59) to all of the Ind of the power set 2^Ω yields:

Given B = {W} and D = {E}, the positive boundary region is defined :

$POS_B(D) = \bigcup_{x \in D} B_*(X)$

$POS_{\{W\}}(E) = \bigcup_{x \in E}\{W\}_*(X) = \bigcup\{\{1,4,5,8\}, \{14\}\} = \{1,4,5,8,14\}$

Applying (6.60) or (6.61) to B and E yields:

$\gamma(B,D) = |\{1,4,5,8,14\}|/9 = 5/9$

Given B = {H} and D = {E}, the positive boundary region is defined:

$POS_{\{H\}}(E) = \bigcup_{x \in E}\{H\}_*(X) = \bigcup\{\{9,20\}, \{\emptyset,\emptyset\}\} = \{9,20\}$

Applying (6.60) or (6.61) to B and E yields:

$\gamma(\{H\}, D) = |\{9,20\}|/20 = 2/9$

Given B = {SE} and D = {E}, the positive boundary region is defined :

$POS_{\{SE\}}(E) = \bigcup_{x \in E}\{SE\}_*(X) = \bigcup\{\emptyset, \emptyset, \emptyset\} = \emptyset$

Applying (6.60) or (6.61) to {B} and {E} yields:

$\gamma(\{SE\}, D) = |\emptyset|/20 = 0$

Because $\gamma(\{SE\}, D) = 0$, thus E is independent of SE.

Given B = {W, H} and D = {E} the positive boundary region is defined:

$POS_{\{W,H\}}(E) = \bigcup_{x \in E}\{W,H\}_*(X) = \bigcup\{\{3,11,14\}, \{1,9,20\}\}$
$= \{1,3,9,11,14,20\}$

Applying (6.60) or (6.61) to B and E yields:

$\gamma(\{W,H\}, D) = |1,3,4,5,8,9,14|/9 = 6/9.$

1 set of attributes B = {W, H} is left. That set has a $\gamma(\{W,H\}) = 6/9$. It
constitutes the minimal reductset K = {{W, H}}.

   Deterministic Rules extracted from Table 6.29 by applying steps a-f of SSP
as shown in tables 6.28 through 6.29 yield:

Table 6.27. Table 6.20 with 2 attributes W and H

Users	W	H	E
3	M	L	SU
11	S	M	SU
14	VL	M	SU
1	L	M	F
4	L	L	F
5	L	L	F
8	L	L	F
9	M	S	F
20	S	S	F

Table 6.28. Applying steps a-b to table 6.27

Users	W	H	E
3	M	L	SU
11	S	M	SU
14	VL	M	SU
1	L	M	F
9	M	S	F
20	S	S	F

Table 6.29. Applying steps e-f to table 6.28

Users	W	H	E
20,9	X	S	F
1	L	X	F
3	X	L	SU
11	X	M	SU
14	VL	X	SU

1. 3 Rules on Success:
    (a) If (W=M and H=L) then User=SU
    (b) If (W=VL) then User= SU
    (c) If (W=S and H=M) then User= SU
2. 3 Rules on Failure:
    (a) If (W=M and H=S) then User= F
    (b) If (W=L) then User= F
    (c) (W=S and H=S) then User=F.

The minimal reduct set yielded exact result to the idea Discernibility Function. These 6 deterministic rules are independent of the number of searches or the attribute SE.

Decision Rules: deterministic and non deterministic
Combined with the 6 deterministic rules that were specified above, 2 other non deterministic rules stand out: 2 Contradictory Rules:

1. If (W=S), (H=S), and (SE=S) then User= SU or F.
2. If (W=S), (H=M), and (SE=S) then User= SU or F.

These 2 contradictory rules also called inconsistent or possible or non-deterministic rules have the same conditions but different decisions, so the proper decision cannot be made by applying this kind of rules. Possible decision rules determine a set of possible decision, which can be made on the basis of given conditions. With every possible decision rule, we will associate a credibility factor of each possible decision suggested by the rule. We propose to define a membership function. Let $\delta(x)$ denote the decision rule associated with object x. We will say that x supports rules $\delta(x)$. Then $C(\delta(x))$ can be denoted by:

$$C(\delta(x)) = 1 \text{ if } \mu_{IX}(x) = 0 \text{ or } 1. \qquad (6.71)$$
$$C(\delta(x)) = \mu_{IX}(x), \text{if } 0 < \mu_{IX}(x) < 1$$

A consistent rule is given a credibility factor of 1, and an inconsistent rule is given a credibility factor smaller than 1 but not equal to 0. The closer it is to one the more credible the rule is. The credibility factor of inconsistent rules in the case of success is $5/7 > .5$ which makes more credible while the failure is $2/7$ which makes it less credible (being closer to 0). In fact the above considerations give rise to the question whether "Success" or "Failure" depends on Searches, hyperlinks, Web Pages. As the above analysis, this is not the case in our example since the inconsistent rules do not allow giving a definite answer. In order to deal with this kind of situations, partial dependency of attributes can be defined. Partial dependency is expressed by the degree to which the dependency among attributes hold and is defined by the ratio of the number of deterministic rules to the total number of rules. In our case it is:
Dependency Ratio = Number of deterministic Rules/Total Number of rules
= 6/8 =.75.

Rough Set applied to Case 2: Searches, Time and Web Pages
According to Agosto [2], the participants in her Web session found Time constraints to be a constant problem when using the Web. They considered time limitations particularly restrictive in Web use for homework purposes, causing them to examine only a few possible Websites before selecting one. As one participant explained, "We don't have that much time when we're doing a report or something to look through everything." They agreed that improved search functions would help to reduce wasted search time by allowing them to zero in on the particular pieces of information in which they were interested. Consider the following example: U={1,2,3,4,5,6,7,8,9,10,11,12,13,14,15,16,17, 18,19,20}= set of 20 users who searched the query, Ω ={Searches, Time, Web Pages}={SE,T,W}, a set of 3 attributes. Since some users did search more

Table 6.30. Summary of All Users

Users	W	T	SE	E
1	L	VL	M	F
2	S	M	S	SU
3	M	L	L	SU
4	L	VL	VL	F
5	L	M	L	F
6	M	L	L	SU
7	S	M	S	F
8	L	VL	M	F
9	M	M	L	F
10	M	M	M	SU
11	S	L	M	SU
12	M	L	M	SU
13	M	VL	L	SU
14	VL	VL	VL	SU
15	S	S	S	SU
16	S	S	S	SU
17	S	S	S	SU
18	S	S	S	SU
19	S	S	S	F
20	S	S	M	F

than others, browsed more than others, scrolled down Web pages more than others, a simple transformation yields a table up to 3 different attributes with a set of values +ranging form: small(S), medium(M), large(L), and very large(VL) (see table 6.30).

$\Omega = \{SE,T,W\}$,

$V_{SE}=\{M,S,L,VL,L,L,S,M,L,M,M,M,L,VL,S,S,S,S,S,M\}$,

$V_T=\{VL,M,L,VL,M,L,M,VL,M,L,L,L,VL,VL,S,S,S,S,S,S\}$,

$V_W=\{L,S,M,L,L,M,S,L,M,M,S,M,M,VL,S,S,S,S,S,S\}$.

In a new UMS, S is represented by $S = (U, \Omega, V, f)$, $f_q(p)$ where $q \in \Omega$ and $p \in PU=\{P-E\}$ is the user's knowledge about the query, and $f_q(p)$ where $q \in \Omega$ and $p = E$ is the expert's evaluation of the query for a given student. E is the decision attribute. Consider the above example but this time: $\Omega = \Omega_u \cup \Omega_e=\{SE,T,W\}$ E, where E= SU(success) or F(failure);

$V_{SE}= \{M,S,L,VL,L,L,S,M,L,M,M,M,L,VL,S,S,S,S,S,M\}$,

$V_T= \{VL,M,L,VL,M,L,M,VL,M,L,L,L,VL,VL,S,S,S,S,S,S\}$,

$V_W= \{L,S,M,L,L,M,S,L,M,M,S,M,M,VL,S,S,S,S,S,S\}$,

$V_E= \{F,SU,SU,F,F,SU,F,F,F,SU,SU,SU,SU,SU,SU,SU,SU,SU,F,F\}$.

Indiscernibility

By applying (6.62) to table 6.30 yields the following:

$Ind(\{W\}) = \{\{2,7,11,15,16,17,18,19,20\}_S, \{3,6,9,10,12,13\}_M,$ $\{1,4,5,8\}_L, \{14\}_{VL}\}$

$\text{Ind}(\{T\}) = \{\{15, 16, 17, 18, 19, 20\}_S, \{2, 5, 7, 9, 10\}_M,$
$\{3, 6, 11, 12\}_L, \{1, 4, 8, 13, 14\}_{VL}\}$
$\text{Ind}(\{SE\}) = \{\{2, 7, 15, 16, 17, 18, 19\}_S, \{1, 8, 10, 11, 12, 20\}_M,$
$\{3, 5, 6, 9, 13\}_L, \{4, 14\}_{VL}\}$

By applying (6.62) to the already calculated Ind of the individual attributes W, H, and SE, we find the followings:

$\text{Ind}(\{W, T\}) = \{\{15, 16, 17, 18, 19, 20\}, \{2, 7\}, \{11\}, \{9, 10\}, \{3, 6, 12\},$
$\{13\}, \{5\}, \{1, 4, 8\}, \{14\}\}$
$\text{Ind}(\{W, SE\}) = \{\{2, 7, 15, 16, 17, 18, 19\}, \{11, 20\}, \{10, 12\}, \{3, 6, 9, 13\},$
$\{1, 8\}, \{5\}, \{4\}, \{14\}\}$
$\text{Ind}(\{T, SE\}) = \{\{15, 16, 17, 18, 19\}, \{20\}, \{2, 7\}, \{10\}, \{5, 9\}, \{11, 12\},$
$\{3, 6\}, \{1, 8\}, \{13\}, \{4, 14\}\}$
$\text{Ind}(\{W, T, SE\}) = \{\{2, 15, 17, 18, 19\}, \{7, 16\}, \{10, 12\}, \{6, 13\}, \{3\}, \{20\},$
$\{1\}, \{8\}, \{9\}, \{5\}, \{4\}, \{11\}, \{14\}\}$

Application of the above definitions of Rough set theory to Table 6.30
In table 6.30, users $\{2,7,15,16,17,18,19\}$ are indiscernible according to the attribute SE=S, users $\{1,4,5,8\}$ are indiscernible for the attribute W=L. For example the attribute W generates 4 sets of users: $\{2,7,11,15,16,17,18,19,20\}_S$, $\{3,6,9,10,12,13\}_M$, $\{1,4,5,8\}_L$, *and* $\{14\}_{VL}$. Because users $\{15,16,17,18\}$ were SU and user $\{19\}$ failed, and are indiscernible to attributes W=S, T=S, and SE=S, then the decision variable for SU or F cannot be characterized by W=S, T=S, and SE=S. Also user $\{2\}$ was SU and user $\{7\}$ failed and are indiscernible to attributes W=S, T=M, and SE=S. Hence users $\{2,15,16,17,18\}$ and $\{7,19\}$ are boundary-line cases. The remaining users: $\{3,6,10,11,12,13,14\}$ have characteristics that enable us to classify them as being SU, while users $\{1,4,5,8,9,20\}$ display characteristics that enable us to classify them as F, and users $\{2,7,15,16,17,18,19\}$ cannot be excluded from being SU or F. Thus the lower approximation of the set of being SU by applying (6.55) is:

$I_*(SU)=\{3,6,10,11,12,13,14\}$

and the upper approximation of being SU by applying (6.56) is:

$I^*(SU)=\{2,3,6,7,10,11,12,13,14,15,16,17,18,19\}$.

Similarly in the concept of F, its lower approximation by applying (6.55) is:

$I_*(F)=\{1,4,5,8,9,20\}$

and its upper approximation by applying (6.56) is:

$I^*(F)=\{1,4,5,8,9,20,2,7,15,16,17,18,19,20\}$.

The boundary region of the set SU or F by applying (6.57) is:

$BN_I(SU) = BN_I(F) = \{2, 7, 15, 16, 17, 18, 19\}$.

The positive region of being SU and F by applying (6.59) is:

$I_*(F) \cup I_*(SU) = \{1, 3, 4, 5, 6, 8, 9, 10, 11, 12, 13, 14, 20\} \neq U$

$I_*(F) \cup I_*(SU)$ is called the I-positive region of S and is denoted by $POS_I(SE)$.

Figure 6.29 shows an equivalence relation of both successful and non successful users using Rough Set Theory. By applying (6.58) to SU:

$\alpha(SU) = |\{3, 6, 10, 11, 12, 13, 14\}|/|\{2, 3, 6, 7, 10, 11, 12, 13, 14, 15, 16, 17, 18, 19\}| = 7/14 = .5$

We also compute the membership value of each user to the concept of SU or F. By applying (6.64) to SU we have:

$\mu SU(1) = |\{2, 3, 6, 10, 11, 12, 13, 14, 15, 16, 17, 18\} \cap \{1\}|/|\{1\}| = 0$

$\mu SU(2) = |\{2, 3, 6, 10, 11, 12, 13, 14, 15, 16, 17, 18\} \cap \{2, 7, 15, 16, 17, 18, 19\}|$
$/|\{2, 7, 15, 16, 17, 18, 19\}|/\{2, 7, 15, 16, 17, 18, 19\} = 5/7$

$\mu SU(3) = |\{2, 3, 6, 10, 11, 12, 13, 14, 15, 16, 17, 18\} \cap \{3\}|/|\{3\}|$
$= |\{3\}|/|\{3\}| = 1$

$\mu SU(4) = |\{2, 3, 6, 10, 11, 12, 13, 14, 15, 16, 17, 18\} \cap \{4\}|/|\{4\}| = 0$

$\mu SU(5) = |\{2, 3, 6, 10, 11, 12, 13, 14, 15, 16, 17, 18\} \cap \{5\}|/|\{5\}| = 0$

$\mu SU(6) = |\{2, 3, 6, 10, 11, 12, 13, 14, 15, 16, 17, 18\} \cap \{6\}|/|\{6\}| = |\{6\}|/$
$|\{6\}| = 1$

$\mu SU(7) = |\{2, 3, 6, 10, 11, 12, 13, 14, 15, 16, 17, 18\} \cap \{2, 7, 15, 16, 17, 18, 19\}|$
$/|\{2, 7, 15, 16, 17, 18, 19\}| = 5/7$

$\mu SU(8) = |\{2, 3, 6, 10, 11, 12, 13, 14, 15, 16, 17, 18\} \cap \{8\}|/|\{8\}| = 0$

$\mu SU(9) = |\{2, 3, 6, 10, 11, 12, 13, 14, 15, 16, 17, 18\} \cap \{9\}/|\{9\}\} = 0$

$\mu SU(10) = |\{2, 3, 6, 10, 11, 12, 13, 14, 15, 16, 17, 18\} \cap \{10\}|/|\{10\}| = |\{10\}$
$|/|\{10\}| = 1$

$\mu SU(11) = |\{2, 3, 6, 10, 11, 12, 13, 14, 15, 16, 17, 18\} \cap \{11\}|/\{11\}| = |\{11\}|/$
$|\{11\}| = 1$

$\mu SU(12) = |\{2, 3, 6, 10, 11, 12, 13, 14, 15, 16, 17, 18\} \cap \{12\}|/|\{12\}| = 1$

$\mu SU(13) = |\{2, 3, 6, 10, 11, 12, 13, 14, 15, 16, 17, 18\} \cap \{13\}|/\{13\}| = 1$

$\mu SU(14) = |\{2, 3, 6, 10, 11, 12, 13, 14, 15, 16, 17, 18\} \cap \{14\}|/|\{14\}| = 1$

$\mu SU(15) = |\{2, 3, 6, 10, 11, 12, 13, 14, 15, 16, 17, 18\} \cap \{2, 7, 15, 16, 17, 18, 19\}|$
$/|\{2, 7, 15, 16, 17, 18, 19\}| = 4/5$

$\mu SU(16) = |\{2, 3, 6, 10, 11, 12, 13, 14, 15, 16, 17, 18\} \cap \{2, 7, 15, 16, 17, 18, 19\}|$
$/\{2, 7, 15, 16, 17, 18, 19\}| = 5/7$

$\mu SU(17) = |\{2, 3, 6, 10, 11, 12, 13, 14, 15, 16, 17, 18\} \cap \{15, 16, 17, 18, 19\}|$
$/|\{15, 16, 17, 18, 19\}| = 5/7$

$\mu SU(18) = |\{2, 3, 6, 10, 11, 12, 13, 14, 15, 16, 17, 18\} \cap \{15, 16, 17, 18, 19\}|/$
$|\{15, 16, 17, 18, 19\}| = 5/7$

$\mu SU(19) = |\{2, 3, 6, 10, 11, 12, 13, 14, 15, 16, 17, 18\} \cap \{15, 16, 17, 18, 19\}|/$
$|\{15, 16, 17, 18, 19\} = 5/7$

$\mu SU(20) = |\{2, 3, 6, 10, 11, 12, 13, 14, 15, 16, 17, 18\} \cap \{20\}| = 0$

$\mu SU = \{0/1, .71/2, 1/3, 0/4, 0/5, 1/6, .71/7, 0/8, 0/9, 1/10, 1/11, 1/12, 1/13,$
$1/14, .71/15, .71/16, .71/17, .71/18, .71/19, 0/20\}$

By applying (6.58) to F:

$\alpha(F) = |\{1, 4, 5, 8, 9, 20\}|/|\{1, 2, 4, 5, 7, 8, 9, 15, 16, 17, 18, 19, 20\}| = 6/14 = .43$

By applying (6.64) we have F as follows:

$\mu F(1) = |\{1, 4, 5, 7, 8, 9, 19, 20\} \cap \{1\}|/|\{1\}| = 1$

$\mu F(2) = |\{1,4,5,7,8,9,19,20\} \cap \{2,7,15,16,17,18,19\}|/|\{2,7,15,16,17,18,19\}| = 2/7$

$\mu F(3) = |\{1,4,5,7,8,9,19,20\} \cap \{3\}|/\{3\}| = 0$

$\mu F(4) = |\{1,4,5,7,8,9,19,20\} \cap \{4\}|/|\{4\}| = 1$

$\mu F(5) = |\{1,4,5,7,8,9,19,20\} \cap \{5\}|/|\{5\}| = 1$

$\mu F(6) = |\{1,4,5,7,8,9,19,20\} \cap \{6\}|/|\{6\}| = 0$

$\mu F(7) = |\{1,4,5,7,8,9,19,20\} \cap \{7\}|/|\{7\}| = 1$

$\mu F(8) = |\{1,4,5,7,8,9,19,20\} \cap \{8\}|/|\{8\}| = 1$

$\mu F(9) = |\{1,4,5,7,8,9,19,20\} \cap \{9\}|/|9| = 1$

$\mu F(10) = |\{1,4,5,7,8,9,19,20\} \cap \{10\}|/|10| = 0$

$\mu F(11) = |\{1,4,5,7,8,9,19,20\} \cap \{11\}|/|11| = 0$

$\mu F(12) = |\{1,4,5,7,8,9,19,20\} \cap \{12\}|/|12| = 0$

$\mu F(13) = |\{1,4,5,7,8,9,19,20\} \cap \{13\}|/|13| = 0$

$\mu F(14) = |\{1,4,5,7,8,9,19,20\} \cap \{14\}|/|14| = 0$

$\mu F(15) = |\{1,4,5,7,8,9,19,20\} \cap \{2,7,15,16,17,18,19\}|/|\{2,715,16,17,18,19\}| = 2/7$

$\mu F(16) = |\{1,4,5,7,8,9,19,20\} \cap \{2,7,15,16,17,18,19\}|/|\{2,7,15,16,17,18,19\}| = 2/7$

$\mu F(17) = |\{1,4,5,7,8,9,19,20\} \cap \{2,7,15,16,17,18,19\}|/|\{2,7,15,16,17,18,19\}| = 2/7$

$\mu F(18) = |\{1,4,5,7,8,9,19,20\} \cap \{2,7,15,16,17,18,19\}|/|\{2,7,15,16,17,18,19\}| = 2/7$

$\mu F(19) = |\{1,4,5,7,8,9,19,20\} \cap \{15,16,17,18,19\}|/|\{2,7,15,16,17,18,19\}| = 2/7$

$\mu F(20) = |\{1,4,5,7,8,9,19,20\} \cap \{20\}|/|\{20\}| = 1$

$\mu F = \{1/1, .29/2, 0/3, 1/4, 1/5, 0/6, .29/7, 1/8, 1/9, 0/10, 0/11, 0/12, 0/13, 0/14, .29/15, .29/16, .29/17, .29/18, .29/19, 1/20\}$

Figure 6.28 shows the Rough membership function of both F and SU

Decision Rules according to Rough Set Theory

Applying the SSP procedure from steps a-b results in table 6.31. The number of users is reduced from 20 to 18 users because of steps a-b of SSP. According to table 6.31, the following 18 rules on Success and Failure can be made by just applying steps a-b:

1. If (W=S), (T=S), and (SE=S) then User= F
2. If (W=S), (T=S), and (SE=M) then User= F
3. If (W=S), (T=L), and (SE=S) then User= F
4. If (W=M), (T=M), and (SE=L) then User= F
5. If (W=L), (T=M), and (SE=L) then User= F
6. If (W=L), (T=VL), and (SE=M) then User= F
7. If (W=L), (T=VL), and (SE=VL) then User= F
8. If (W=VL), (T=VL), and (SE=VL) then User=SU
9. If (W=S), (T=S), and (SE=S) then User= SU
10. If (W=S), (T=M), and (SE=S) then User= SU

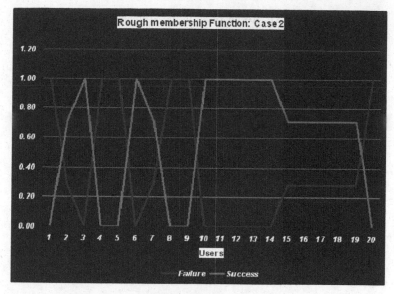

Fig. 6.28. Case 2

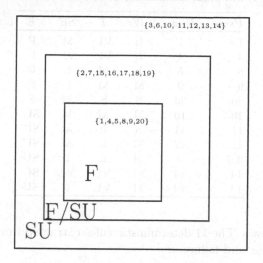

Fig. 6.29. Approximating SU and F using W, T, and SE

11. If (W=S), (T=L), and (SE=M) then User= SU
12. If (W=M), (T=L), and (SE=M) then User= SU
13. If (W=M), (T=L), and (SE=L) then User= SU
14. If (W=M), (T=VL), and (SE=L) then User= SU

Since users {15,16,17,18} and {19} and users {2} and {7} are indiscernible they can be made out one contradictory rule and reduce the number of rows

Table 6.31. Result of applying steps a-b of SSP

Users	Users	W	T	SE	E
1,8	1	L	VL	M	F
4	4	L	VL	VL	F
5	5	L	M	L	F
7	7	S	M	S	F
9	9	M	M	L	F
19	19	S	S	S	F
20	20	S	S	M	F
10	10	M	M	M	SU
11	11	S	L	M	SU
12	12	M	L	M	SU
3,6	3	M	L	L	SU
14	14	VL	VL	VL	SU
2	2	S	M	S	SU
15,16,17,18	17	S	S	S	SU
13	13	M	VL	L	SU

Table 6.32. Eliminating all contradictory users

Users	Users	W	T	SE	E
1,8	1	L	VL	M	F
4	4	L	VL	VL	F
5	5	L	M	L	F
9	9	M	M	L	F
20	20	S	S	M	F
10	10	M	M	M	SU
11	11	S	L	M	SU
12	12	M	L	M	SU
3,6	3	M	L	L	SU
14	14	VL	VL	VL	SU
13	13	M	VL	L	SU

from 18 to 11 rows. The 11 deterministic rules extracted from table 6.32 as it applies to success and failure are:

1. If (W=S), (T=S), and (SE=M) then User= F
2. If (W=S), (T=L), and (SE=S) then User= F
3. If (W=M), (T=M), and (SE=L) then User= F
4. If (W=L), (T=M), and (SE=L) then User= F
5. If (W=L), (T=VL), and (SE=M) then User= F
6. If (W=L), (T=VL), and (SE=VL) then User= F
7. If (W=VL), (T=VL), and (SE=VL) then User=SU
8. If (W=S), (T=M), and (SE=S) then User= SU
9. If (W=S), (T=L), and (SE=M) then User= SU

Table 6.33. 9 users are left after applying steps a-d of SSP to table 6.32

Users	Users	W	T	SE	E
4	4	L	VL	VL	F
5	5	L	M	L	F
9	9	M	M	L	F
20	20	S	S	M	F
10	10	M	M	M	SU
11	11	S	L	M	SU
12	12	M	L	M	SU
14	14	VL	VL	VL	SU
13	13	M	VL	L	SU

10. If (W=M), (T=L), and (SE=M) then User= SU
11. If (W=M), (T=L), and (SE=L) then User= SU
12. If (W=M), (T=VL), and (SE=L) then User= SU

By eliminating all identical examples we have nine users left out of the original 20 users:

The 9 deterministic rules extracted from table 6.33 as it applies to success and failure are:

1. If (W=S), (T=S), and (SE=M) then User= F
2. If (W=M), (T=M), and (SE=L) then User= F
3. If (W=L), (T=M), and (SE=L) then User= F
4. If (W=L), (T=VL), and (SE=VL) then User= F
5. If (W=M), (T=M), and (SE=M) then User= SU
6. If (W=M), (T=L), and (SE=L) then User= SU
7. If (W=VL), (T=VL), and (SE=VL) then User=SU
8. If (W=S), (T=L), and (SE=M) then User= SU
9. If (W=M), (T=VL), and (SE=L) then User= SU

Discernibiliy Matrix and Discernibility Function
Example: Applying (6.59) to table 6.33 yields the following Discernibility Function:

$$f_A = \{\{W,T,SE\}, \{W,T,SE\}, \{W,T,SE\}, \{W,SE\}, \{W\}, \{W,SE\},$$
$$\{W,T,SE\}, \{W,T,SE\}, \{W,T\}, \{W,T,SE\}, \{SE\}, \{W,T,SE\}, \{T,SE\},$$
$$\{T\}, \{W,T,SE\}, \{W,T\}, \{T\}, \{W,T\}, \{W,T,SE\}, \{W,T,SE\}\}$$
$$f_A = \{\{W\}, \{\{W,T\}, \{W,SE\}\}, \{\{SE\}, \{T\}\}, \{T\}\}$$
$$f_A = \{\{W\}, \{SE\}, \{T\}\}$$

Applying (6.58) yields a symmetric matrix (9,9) as seen below which is half filled. The result of applying of steps e through f and knowing that the 3 attributes are needed are shown in table 6.34 and table 6.35.

Table 6.34. Core of the set of final data (Step e- of SSP)

Users	Users	W	T	SE	E
4	4	L	X	X	F
5	5	L	X	X	F
9	9	X	M	L	F
20	20	S	S	M	F
10	10	X	M	M	SU
11	11	X	L	X	SU
12	12	X	L	X	SU
14	14	VL	X	X	SU
13	13	X	VL	L	SU

Table 6.35. Set of reduct set (Step f- of SSP)

Users	Users	W	T	SE	E
4,5	4	L	X	X	F
9	9	X	M	L	F
20	20	X	S	M	F
10	10	X	M	M	SU
11,12	11	X	L	X	SU
14	14	VL	X	X	SU
13	13	X	VL	L	SU

$$
\begin{array}{l}
\quad\quad\ 4 \quad\quad\quad 5 \quad\quad\quad\quad 9 \quad\quad\quad\quad 20 \quad\ 10\ \ 11\ \ 12\ \ 13\ \ 14 \\
\begin{array}{l}
4 \\ 5 \\ 9 \\ 20 \\ 10 \\ 11 \\ 12 \\ 13 \\ 14
\end{array}
\left[
\begin{array}{lllllllll}
\emptyset & & & & & & & & \\
\emptyset & \emptyset & & & & & & & \\
\emptyset & \emptyset & \emptyset & & & & & & \\
\emptyset & \emptyset & \emptyset & \emptyset & & & & & \\
W,T,SE & W,SE & SE & W,T & \emptyset & & & & \\
W,T,SE & W,T,SE & W,T,SE & T & \emptyset & \emptyset & & & \\
W,T,SE & W,T,SE & T,SE & W,T & \emptyset & \emptyset & \emptyset & & \\
W,SE & W,T & T & W,T,SE & \emptyset & \emptyset & \emptyset & \emptyset & \\
W & W,T,SE & W,T,SE & W,T,SE & \emptyset & \emptyset & \emptyset & \emptyset & \emptyset
\end{array}
\right]
\end{array}
$$

1. 4 Rules on Success:
   (a) If (T=M and SE=M) then User= SU
   (b) If (T=VL and SE=L) the User=SU
   (c) If (T=L) then User=SU
   (d) If (W=VL) then User= SU

2. 3 Rules on Failure:
   (a) If (T=M and SE=L) then User= F
   (b) If (T=S) and SE=M) the user =F
   (c) If (W=L) then User= F

These 7 rules say that the 3 parameters are needed to classify the 8 determin-istic examples or users out of the original 20.

Reducts, reduct Sets(RS), and Minimum reduct Set(MRS)

In our example $\Omega = \{W,T,SE\}$ and D=$\{S\}$, and thus the power set of Ω is equal to: $2^{\Omega} = \{\emptyset,\{W\},\{T\},\{SE\},\{W,T\},\{W,SE\},\{T,SE\},\{W,T,SE\}\}$ K is said to be a minmimal RS or MRS iff it is a reductset RS with the minimum number of elements in it such that:

$$|K| \leq |R|\forall R \in 2^{\Omega}$$

Applying the concept of minimal reduct set to find deterministic rules

The lower approximation of the set of being SU by applying (6.55) to table 6.33 is:

$I_*(SU)=\{10,11,12,13,14\}$

and the upper approximation of being SU by applying (6.56) is:

$I^*(SU)=\{10,11,12,13,14\}$.

Similarly in the concept of F, its lower approximation by applying (6.55) is:

$I_*(F)=\{4,5,9,20\}$

and its upper approximation by applying (6.56) is:

$I^*(F)=\{4,5,9,20\}$.

Table 6.33 has no boundary regions.

By applying (6.62) and (6.63) to table 6.33 yields the following:

$Ind(\{W\}) = \{\{11,20\}_S, \{9,10,12,13\}_M, \{4,5\}_L, \{14\}_{VL}\}$
$Ind(\{SE\}) = \{\{10,11,12,20\}_M, \{5,9,13\}_L, \{4,14\}_{VL}\}$
$Ind(\{T\}) = \{\{20\}_S, \{5,9,10\}_M, \{11,12\}_L, \{4,13,14\}_{VL}\}$
$Ind(\{W,T\}) = \{\{20\}, \{11\}, \{9,10\}, \{12\}, \{13\}, \{4\}, \{5\}, \{14\}\}$.
$Ind(\{W,SE\}) = \{\{11,20\}, \{10,12\}, \{9,13\}, \{5\}, \{4\}, \{14\}\}$
$Ind(\{SE,T\}) = \{\{20\}, \{10\}, \{11,12\}, \{5,9\}, \{13\}, \{4,14\}\}$
$Ind(\{W,T,SE\}) = \{\{3\}, \{20\}, \{1\}, \{8\}, \{9\}, \{5\}, \{4\}, \{11\}, \{14\}\}$

Applying (6.59) to all of the Ind of the power set 2^{Ω} yields:
Given B=$\{W\}$ and D=$\{E\}$, the positive boundary region is defined:

$POS_B(D) = \bigcup_{x \in D}(B_*(X))$
$POS_{\{W\}}(E) = \bigcup_{x \in E}(\{W\}_*)(X) = \bigcup\{\{4,5\},\{14\}\} = \{4,5,14\}$

Applying (6.60) or (6.61) to B and E yields:

$\gamma(B,D) = |\{4,5,14\}|/8 = 3/8$
Given $B = \{SE\}$ and $D = \{E\}$, the positive boundary region is defined :

$POS_{\{H\}}(E) = \bigcup_{x \in \{E\}}\{SE\}_*(X) = \bigcup\{\emptyset,\emptyset\} = \emptyset$

Applying (6.60) or (6.61) to B and E yields:

$\gamma(\{SE\},D) = |\emptyset|/9 = 0/9$

Given $B = \{T\}$ and $D = \{E\}$, the positive boundary region is defined :

$POS_{\{SE\}}(E) = \bigcup_{x \in \{E\}} \{T\}_*(X) = \bigcup \{\{20\}, \emptyset\} = \{20\}$

Applying (6.60) or (6.61) to B and E yields:

$\gamma(\{T\}, D) = |\{20\}|/8 = 1/8$

Given $B = \{W, T\}$ and $D = \{E\}$ the positive boundary region is defined:

$POS_{\{W,H\}}(E) = \bigcup_{x \in \{E\}} \{W, H\}_*(X) =$
$\bigcup \{\{11, 12, 13, 14\}, \{4, 5, 20\}\} = \{4, 5, 11, 12, 13, 14, 20\}$

Applying (6.60) or (6.61) to B and E yields:

$\gamma(\{W, T\}, D = |\{4, 5, 9, 11, 12, 13, 14, 20\}|/9 = 7/9.$

Given $B = \{W, SE\}$ and $D = \{E\}$ the positive boundary region is defined:

$POS_{\{W,SE\}}(E) = \bigcup_{x \in \{E\}} \{W, SE\}_*(X) = \bigcup \{\{14\}, \{4, 5\}\} = \{4, 5, 14\}$

Applying (6.60) or (6.61) to B and E yields:

$\gamma(\{W, SE\}, D) = |\{4, 5, 7, 14\}|/9 = 3/9.$

Given $B = \{SE, T\}$ and $D = \{E\}$ the positive boundary region is defined:

$POS_{\{SE,T\}}(E) = \bigcup_{x \in \{E\}} \{SE, T\}_*(X) = \bigcup \{\{11, 13\}, \{20\}\} = \{11, 13, 20\}$

Applying (6.60) or (6.61) to B and E yields:

$\gamma(\{SE, T\}, D) = |\{7, 11, 13, 20\}|/9 = 3/9.$

set of B's: $B = \{W, SE\}$ and $B = \{SE, T\}$ have the lowest $\gamma(\{B\}, D)$. That set has $\gamma(\{B\}, D) = 3/9$, It constitutes the minimal reductset $K = \{\{WSE\}, \{SET\}\}$.
Deterministic rules extracted from table 6.32 by applying steps a-f of SSP (see table 6.36 and table 6.37) are:

1. 4 Rules on Success:

   (a) If (T=M and SE=M) then User= SU
   (b) If (T=VL and SE=L) the User=SU

Table 6.36. Core Set of table 6.32

Users	Users	W	T	SE	E
4	4	L	X	X	F
5	5	L	X	X	F
9	9	X	X	L	F
20	20	X	X	M	F
10	10	X	M	X	SU
11	11	X	L	X	SU
12	12	X	L	X	SU
14	14	VL	X	X	SU
13	13	X	VL	X	SU

Table 6.37. Reduct Set of table 6.32

Users	Users	W	T	SE	E
4,5	4	L	X	X	F
9	9	X	X	L	F
20	20	X	X	M	F
10	10	X	M	X	SU
11	11	X	L	X	SU
12	12	X	L	X	SU
14	14	VL	X	X	SU
13	13	X	VL	X	SU

   (c) If (T=L) then User=SU

   (d) If (W=VL) then User= SU

2. 3 Rules on Failure:

   (a) If (T=M and SE=L) then User= F

   (b) If (T=S and SE=M) then User =F

   (c) If (W=L) then User= F

Decision Rules: Deterministic and non Deterministic

Combined with the 7 deterministic rules that were specified above, 1 other non deterministic rule stand out: 2 Contradictory Rules:

1. If (W=S), (H=S), and (SE=S) then User= SU or F.
2. If (W=S), (H=M), and (SE=S) then User= SU or F.

Contradictory rules also called inconsistent or possible or non deterministic rules have the same conditions but different decisions, so the proper decision cannot be made by applying this kind of rules. Possible decision rules determine a set of possible decision, which can be made on the basis of given conditions. With every possible decision rule, we will associate a credibility factor of each possible decision suggested by the rule. A consistent rule is given a credibility factor of 1, and an inconsistent rule is given a credibility factor smaller than 1 but not equal to 0. The closer it is to one the more credible the rule is. By applying (6.70), the credibility factor of inconsistent rules in the case of Success is 5/7 >.5 which makes it very credible, while it is 2/7 in the case of Failure which makes less credible (being closer to 0). In fact the above considerations give rise to the question whether "Success" or "Failure" depends on Searches, Time, Web Pages. As the above analysis, this is not the case in our example since the inconsistent rules do not allow giving a definite answer. In order to deal with this kind of situations, partial dependency of attributes can be defined. Partial dependency is expressed by the degree to which the dependency among attributes holds and is defined by the ratio of the number of deterministic rules to the total number of rules.

In our case it is: Dependency Ratio = Number of deterministic rules/Total Number of rules = 7/9 =.78

Rough Set applied to Case 3: WebPages, hyperlinks, Searches, and Time Let us illustrate how a decision table such as table 6.38 defines an indiscernibility relation. By applying (6.62) to table 6.38 yields the following:

$Ind(\{W\}) = \{\{2, 7, 11, 15, 16, 17, 18, 19, 20\}_S, \{3, 6, 9, 10, 12, 13\}_M,$
$\{1, 4, 5, 8\}_L, \{14\}_{VL}\}$
$Ind(\{T\}) = \{\{15, 16, 17, 18, 19, 20\}_S, \{2, 5, 7, 9, 10\}_M, \{3, 6, 11, 12\}_L,$
$\{1, 4, 8, 13, 14\}_{VL}$
$Ind(\{SE\}) = \{\{2, 7, 15, 16, 17, 18, 19\}_S, \{1, 8, 10, 11, 12, 20\}_M, \{3, 5, 6, 9, 13\}$
$_L, \{4, 14\}_{VL}$
$Ind(\{H\}) = \{\{2, 9, 10, 12, 15, 17, 18, 19, 20\}_S, \{1, 6, 7, 11, 13, 14, 16\}_M,$
$\{3, 4, 5, 8\}_L$

By applying (6.63) to the already calculated Ind of the individual attributes W, H, T, and SE, we find:
$Ind\{W, H\} = \{\{2, 15, 17, 18, 19, 20\}, \{7, 11, 16\}, \{9, 10, 12\}, \{6, 13\}, \{3\},$
$\{4, 5, 8\}, \{1\}, \{14\}\}$
$Ind\{W, SE\} = \{\{2, 7, 15, 16, 17, 18, 19\}, \{11, 20\}, \{10, 12\}, \{3, 6, 9, 13\}, \{1, 8\},$
$\{5\}, \{4\}, \{14\}\}$
$Ind\{H, SE\} = \{\{2, 15, 17, 18, 19\}, \{10, 12, 20\}, \{7, 16\}, \{1, 11\}, \{6, 13\}, \{8\},$
$\{9\}, \{3, 5\}, \{4\}, \{14\}\}$
$Ind\{W, T\} = \{\{17, 18, 19, 20\}, \{2, 15, 16\}, \{7, 11\}, \{9\}, \{3, 6, 10, 12\}, \{13\},$
$\{5\}, \{1, 4, 8\}, \{14\}\}$

Table 6.38. Table with all 4 attributes: W, H, SE, and T

Users	W	H	SE	T	E
1	L	M	M	VL	F
2	S	S	S	M	SU
3	M	L	L	L	SU
4	L	L	VL	VL	F
5	L	L	L	M	F
6	M	M	L	L	SU
7	S	M	S	M	F
8	L	L	M	VL	F
9	M	S	L	M	F
10	M	S	M	M	SU
11	S	M	M	L	SU
12	M	S	M	L	SU
13	M	M	L	VL	SU
14	VL	M	VL	VL	SU
15	S	S	S	S	SU
16	S	M	S	S	SU
17	S	S	S	S	SU
18	S	S	S	S	SU
19	S	S	S	S	F
20	S	S	M	S	F

$\text{Ind}\{T, SE\} = \{\{17, 18, 19\}, \{20\}, \{2, 15, 16\}, \{5, 9\}, \{7\}, \{10, 11, 12\}, \{3, 6\},$
$\{1, 8\}, \{13\}, \{4, 14\}\}$

$\text{Ind}\{H, T\} = \{\{15, 17, 18, 19, 12\}, \{16\}, \{2, 9, 10\}, \{7\}, \{5\}, \{12\}, \{6, 11\}, \{3\},$
$\{1, 13, 14\}, \{4, 8\}\}$

$\text{Ind}\{W, T, SE\} = \{\{2, 15, 17, 18, 19\}, \{7, 16\}, \{10, 12\}, \{6, 13\}, \{3\}, \{20\}, \{1\},$
$\{8\}, \{9\}, \{5\}, \{4\}, \{11\}, \{14\}\}$

$\text{Ind}\{W, H, SE\} = \{\{2, 15, 17, 18, 19\}, \{20\}, \{7, 16\}, \{11\}, \{10, 12\}, \{9\}, \{6, 13\},$
$\{3\}, \{8\}, \{5\}, \{4\}, \{1\}, \{14\}\}$

$\text{Ind}\{W, T, H\} = \{\{17, 18, 19, 20\}, \{2, 15\}, \{16\}, \{7, 11\}, \{9\}, \{10, 12\}, \{6\}, \{3\},$
$\{13\}, \{5\}, \{1\}, \{4, 8\}, \{14\}\}$

$\text{Ind}\{H, SE, T\} = \{\{15, 17, 18, 19\}, \{2\}, \{20\}, \{10\}, \{12\}, \{16\}, \{7\}, \{11\}, \{1\},$
$\{6\}, \{13\}, \{8\}, \{9\}, \{5\}, \{3\}, \{4\}, \{14\}\}$

Application of the above definitions of Rough set theory to Table 6.38

In table 6.38, users $\{2,7,15,16,17,18,19\}$ are indiscernible according to the attribute SE=S, users $\{1,4,5,8\}$ are indiscernible for the attribute W=L. For example the attribute W generates 4 sets of users:
$\{2,7,11,15,16,17,18,19,20\}_S$, $\{3,6,9,10,12,13\}_M$, $\{1,4,5,8\}_L$, and $\{14\}_{VL}$.
Because users $\{15,17,18\}$ were SU and users $\{19\}$ failed, and are indiscernible to attributes W=S, T=S, and SE=S, then the decision variable for SU or F cannot be characterized by W=S, T=S, and SE=S. Hence users $\{15,17,18\}$ and $\{19\}$ are boundary-line cases. The remaining users: $\{2,3,6,10,11,12,13,14,16\}$ have characteristics that enable us to classify them as being SU, while users $\{1,4,5,7,8,9,20\}$ display characteristics that enable us to classify them as F, and users $\{15,17,18,19\}$ cannot be excluded from being SU or F. Thus the lower approximation of the set of being SU by applying (6.55) is:
$I_*(SU)=\{2,3,6,10,11,12,13,14,16\}$
and the upper approximation of being SU by applying (6.56) is:
$I^*(SU)=\{2,3,6,10,11,12,13,14,16,15,17,18,19\}.$
Similarly in the concept of F, its lower approximation by applying (6.55) is:
$I_*(F)=\{1,4,5,7,8,9,20\}$
and its upper approximation by applying (6.56) is:
$I^*(F)=\{1,4,5,7,8,9,20,15,17,18,19\}.$
The boundary region of the set SU or F by applying (6.57) is:
$BN_I(SU)=BN_I(F)=\{15,17,18,19\}.$
The positive region of being SU and F by applying (6.59) is:
$I_*(F) \cap I_*(SU) = \{1, 2, 3, 4, 5, 6, 7, 8, 9, 10, 11, 12, 13, 14, 16, 20\} \neq U$
$I_*(F) \cap I_*(SU)$ is called the I-positive region of S and is denoted by $POS_I(SE)$.
Figure 6.31 shows an equivalence relation of both successful and non successful users using Rough Set Theory. By applying (6.58) to SU:
$\alpha(SU) = |\{2, 3, 6, 10, 11, 12, 13, 14, 16\}|/|\{2, 3, 6, 10, 11, 12, 13, 14, 16, 15, 17, 18, 19\}| = 9/13 = .7$

We also compute the membership value of each user to the concept of "SU" or "F".

By applying (6.64) to SU we have:

$\mu SU(1) = |\{2,3,6,10,11,12,13,14,16,15,17,18\} \cap \{1\}|/|\{1\}| = 0$

$\mu SU(2) = |\{2,3,6,10,11,12,13,14,16,15,17,18\} \cap \{2\}|/|\{2\}|/\{2\} = 1$

$\mu SU(3) = |\{2,3,6,10,11,12,13,14,16,15,17,18\} \cap \{3\}|/|\{3\}| = |\{3\}|/\{3\}| = 1$

$\mu SU(4) = |\{2,3,6,10,11,12,13,14,15,16,17,18\} \cap \{4\}|/|\{4\}| = 0$

$\mu SU(5) = |\{2,3,6,10,11,12,13,14,15,16,17,18\} \cap \{5\}|/|\{5\}| = 0$

$\mu SU(6) = |\{2,3,6,10,11,12,13,14,15,16,17,18\} \cap \{6\}|/|\{6\}| = |\{6\}|/|\{6\}| = 1$

$\mu SU(7) = |\{2,3,6,10,11,12,13,14,15,16,17,18\} \cap \{7\}|/|\{7\}| = 0$

$\mu SU(8) = |\{2,3,6,10,11,12,13,14,15,16,17,18\} \cap \{8\}|/|\{8\}| = 0$

$\mu SU(9) = |\{2,3,6,10,11,12,13,14,15,16,17,18\} \cap \{9\}/|\{9\}\} = 0$

$\mu SU(10) = |\{2,3,6,10,11,12,13,14,15,16,17,18\} \cap \{10\}|/|\{10\}| =$
$|\{10\}|/|\{10\}| = 1$

$\mu SU(11) = |\{2,3,6,10,11,12,13,14,15,16,17,18\} \cap \{11\}|/|\{11\}| =$
$|\{11\}|/|\{11\}| = 1$

$\mu SU(12) = |\{2,3,6,10,11,12,13,14,15,16,17,18\} \cap \{12\}|/|\{12\}| = 1$

$\mu SU(13) = |\{2,3,6,10,11,12,13,14,15,16,17,18\} \cap \{13\}|/|\{13\}| = 1$

$\mu SU(14) = |\{2,3,6,10,11,12,13,14,16,15,17,18\} \cap \{14\}|/|\{14\}| = 1$

$\mu SU(15) = |\{2,3,6,10,11,12,13,14,16,15,17,18\} \cap \{15,17,18,19\}|/$
$|\{15,17,18,19\}| = 3/5 = .75$

$\mu SU(16) = |\{2,3,6,10,11,12,13,14,15,16,17,18\} \cap \{16\}|/|\{16\}| = 1$

$\mu SU(17) = |\{2,3,6,10,11,12,13,14,15,16,17,18\} \cap \{15,17,18,19\}|/|$
$\{15,17,18,19\} = 3/4 = .75$

$\mu SU(18) = |\{2,3,6,10,11,12,13,14,15,16,17,18\} \cap \{15,17,18,19\}|/|$
$\{15,17,18,19\}| = 3/4 = .75$

$\mu SU(19) = |\{2,3,6,10,11,12,13,14,16,15,17,18\} \cap \{15,17,18,19\}|/|$
$\{15,17,18,19\} = 3/4 = .75$

$\mu SU(20) = |\{2,3,6,10,11,12,13,14,15,16,17,18\} \cap \{20\}| = 0$

Thus:

$\mu SU = \{0/1, .1/2, 1/3, 0/4, 0/5, 1/6, .0/7, 0/8, 0/9, 1/10, 1/11, 1/12, 1/13, 1/14,$
$.75/15, 1/16, .75/17, .75/18, .75/19, 0/20\}$

By applying (6.58) to F:

$\alpha(F) = |\{1,4,5,7,8,9,20\}|/|\{1,4,5,7,8,9,20,15,17,18,19\}| = 7/11 = .64$

By applying (6.64) we have F as follows:

$\mu F(1) = |\{1,4,5,7,8,9,19,20\} \cap \{1\}|/|\{1\}| = 1$

$\mu F(2) = |\{1,4,5,7,8,9,19,20\} \cap \{2\}|/|\{2\}| = 0$

$\mu F(3) = |\{1,4,5,7,8,9,19,20\} \cap \{3\}|/|\{3\}| = 0$

$\mu F(4) = |\{1,4,5,7,8,9,19,20\} \cap \{4\}|/|\{4\}| = 1$

$\mu F(5) = |\{1,4,5,7,8,9,19,20\} \cap \{5\}|/|\{5\}| = 1$

$\mu F(6) = |\{1,4,5,7,8,9,19,20\} \cap \{6\}|/|\{6\}| = 0$

$\mu F(7) = |\{1,4,5,7,8,9,19,20\} \cap \{7\}|/|\{7\}| = 1$

$\mu F(8) = |\{1,4,5,7,8,9,19,20\} \cap \{8\}|/|\{8\}| = 1$

$\mu F(9) = |\{1,4,5,7,8,9,19,20\} \cap \{9\}|/|\{9\}| = 1$

$\mu F(10) = |\{1, 4, 5, 7, 8, 9, 19, 20\} \cap \{10\}|/|\{10\}| = 0$
$\mu F(11) = |\{1, 4, 5, 7, 8, 9, 19, 20\} \cap \{11\}|/|\{11\}| = 0$
$\mu F(12) = |\{1, 4, 5, 7, 8, 9, 19, 20\} \cap \{12\}|/|\{12\}| = 0$
$\mu F(13) = |\{1, 4, 5, 7, 8, 9, 19, 20\} \cap \{13\}|/|\{13\}| = 0$
$\mu F(14) = |\{1, 4, 5, 7, 8, 9, 19, 20\} \cap \{14\}|/|\{14\}| = 0$
$\mu F(15) = |\{1, 4, 5, 7, 8, 9, 19, 20\} \cap \{15, 17, 18, 19\}|/|\{15, 17, 18, 19\}| = 1/4.25$
$\mu F(16) = |\{1, 4, 5, 7, 8, 9, 19, 20\} \cap \{16\}|/|\{16\}| = 0$
$\mu F(17) = |\{1, 4, 5, 7, 8, 9, 19, 20\} \cap \{15, 17, 18, 19\}|/|\{15, 17, 18, 19\}| = 1/4$
$\mu F(18) = |\{1, 4, 5, 7, 8, 9, 19, 20\} \cap \{15, 17, 18, 19\}|/|\{15, 17, 18, 19\}| = 1/4$
$\mu F(19) = |\{1, 4, 5, 7, 8, 9, 19, 20\} \cap \{15, 17, 18, 19\}|/|\{15, 17, 18, 19\}| = 1/4$
$\mu F(20) = |\{1, 4, 5, 7, 8, 9, 19, 20\} \cap \{20\}|/|\{20\}| = 1$
$\mu F = \{1/1, 0/2, 0/3, 1/4, 1/5, 0/6, 1/7, 1/8, 1/9, 0/10, 0/11, 0/12, 0/13, 0/14,$
$.25/15, 0/16, .25/17, .25/18, ..25/19, 1/20\}$

Figure 6.30 shows the Rough membership function of both F and SU.

Decision Rules according to Rough Set Theory

Applying the SSP procedure from steps a-b results in table 6.39. Figure 6.30 displays the Rough membership function of both success and failure. The number of users is reduced from 20 to 18 users because of steps a-b of SSP. According to table 6.39, the following 18 rules on Success and Failure can be made by just applying steps a-b:

1. If (W=S), (H=S), (SE=M) and (T=S) then User= F
2. If (W=S), (H=S), (SE=S) and (T=S) then User= F

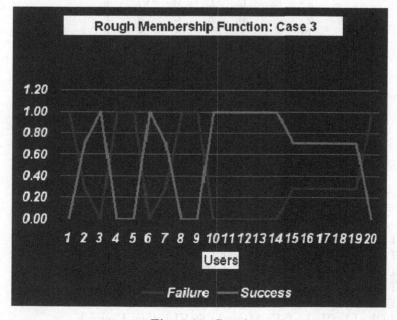

Fig. 6.30. Case 3

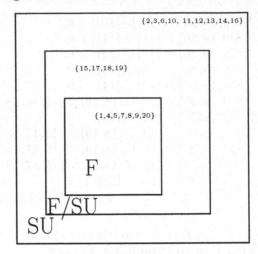

Fig. 6.31. Approximating SU and F using W, T, H, and SE

Table 6.39. Result of applying steps a-b of SSP

Users	W	H	SE	T	E
1	L	M	M	VL	F
4	L	L	VL	VL	F
5	L	L	L	M	F
7	S	M	S	M	F
8	L	L	M	VL	F
9	M	S	L	M	F
19	S	S	S	S	F
20	S	S	M	S	F
2	S	S	S	M	SU
3	M	L	L	L	SU
6	M	M	L	L	SU
10	M	S	M	M	SU
11	S	M	M	L	SU
12	M	S	M	L	SU
13	M	M	L	VL	SU
14	VL	M	VL	VL	SU
15,17,18	S	S	S	S	SU
16	S	M	S	S	SU

3. If (W=S), (H=M), (SE=S) and (T=M), then User= F
4. If (W=M), (H=S) , (SE=L) and (T=M), then User= F
5. If (W=L), (H=L) (SE=L) and (T=M), then User= F
6. If (W=L), (H=M) (SE=M) and (T=VL), then User= F
7. If (W=L), (H=L) (SE=VL) and (T=VL), then User= F
8. If (W=L), (H=L) (SE=M) and (T=VL), then User= F

9. If (W=S), (H=S), (SE=S) and (T=M) then User= SU
10. If (W=VL), (H=M) (SE=VL) and (T=VL), then User=SU
11. If (W=S), (H=M), (SE=M) and (T=L), then User= SU
12. If (W=S), (H=M) (SE=S) and (T=S), then User= SU
13. If (W=M), (H=M) (SE=L) and (T=L), then User= SU
14. If (W=M), (H=L) (SE=L) and (T=L), then User= SU
15. If (W=M), (H=S) (SE= M) and (T=M), then User= SU
16. If (W=M), (H=S) (SE=M) and (T=L), then User= SU
17. If (W=M), (H=M) (SE=L) and (T=VL), then User= SU
18. If (W=S), (H=S), (SE=S) and (T=S)then User= SU

Since users {15,17,18} and {19} are indiscernible they can be made out into a contradictory set of rules and reduce the number of rows from 18 to 16 rows. The 16 deterministic rules extracted from table 6.40 as it applies to success and failure are:

1. If (W=S), (H=S), (SE=M) and (T=S) then User= F
2. If (W=S), (H=M), (SE=S) and (T=M), then User= F
3. If (W=M), (H=S) , (SE=L) and (T=M), then User= F
4. If (W=L), (H=L) (SE=L) and (T=M), then User= F
5. If (W=L), (H=M) (SE=M) and (T=VL), then User= F
6. If (W=L), (H=L) (SE=VL) and (T=VL), then User= F
7. If (W=L), (H=L) (SE=M) and (T=VL), then User= F
8. If (W=S), (H=S), (SE=S) and (T=M) then User= SU
9. If (W=VL), (H=M) (SE=VL) and (T=VL), then User=SU
10. If (W=S), (H=M), (SE=M) and (T=L), then User= SU
11. If (W=S), (H=M) (SE=S) and (T=S), then User= SU

Table 6.40. Eliminating contradictory rules in Table 6.39

Users	W	H	SE	T	E
1	L	M	M	VL	F
4	L	L	VL	VL	F
5	L	L	L	M	F
7	S	M	S	M	F
8	L	L	M	VL	F
9	M	S	L	M	F
20	S	S	M	S	F
2	S	S	S	M	SU
3	M	L	L	L	SU
6	M	M	L	L	SU
10	M	S	M	M	SU
11	S	M	M	L	SU
12	M	S	M	L	SU
13	M	M	L	VL	SU
14	VL	M	VL	VL	SU
16	S	M	S	S	SU

12. If (W=M), (H=M) (SE=L) and (T=L), then User= SU
13. If (W=M), (H=L) (SE=L) and (T=L), then User= SU
14. If (W=M), (H=S) (SE= M) and (T=M), then User= SU
15. If (W=M), (H=S) (SE=M) and (T=L), then User= SU
16. If (W=M), (H=M) (SE=L) and (T=VL), then User= SU

Discernibiliy Matrix and Discernibility Function
Example. Applying (6.70) to Table 6.40 yields the following Discernibility
Function:
f_A ={W,H,SE,T,W,H,SE,T},{W,SE,T},{W,H,T},{W,T},{W,H,T},
{W,SE},{W,SE},{W,SE,T} {W,H,SE,T},{W,SE,T},{W,H,SE,T},
{W,H,SE,T},{W,H,SE,T},{W,H,SE,T},W,H,SE},{W,H},{W,H,SE,T,
{W,H,SE,T},{W,T},{W,H,T},{W,H,SE},{W,H,SE,T},{W,H,SE,T}
,{W,H,T},{W,H,SE,T},{W,H,SE,T},{H},{W,H,SE,T},{W,SE,T},
{W,H,SE},{SE,T},{W,H,SE,T},{W,SE,T},{W,SE,T},{T},
{W,H,SE,T},{W,SE,T}, {W,H,SE,T},{W,H,T},{W,H,T},{W,H,T}
,{W,H,SE},{W,H,SE},{W,H,SE,T},{W,SE},{H,T},{H,T},{SE}
,{W,H,SE,T},{SE,T},{H,T},{W,H,SE,T},{W,H,SE,T},{SE,T},{W,H,SE,T},
{W,H,SE,T},{W,T},{H,T},{W,T},{W,H,SE,T},{W,H,SE,T},{H,SE}
f_A= {{W,T},{W,SE}},{{W,SE,T},{W,H}}{{W,T},{W,H,SE}}{{H},
{T}},{{W,H,SE},{W,H,T}, {W,SE,T}}{{H,T},{SE}}{{SE,T},{W,T}
,{H,T},{H,SE}}= {{W,SE},{SE,T},{W,H}}

Applying (6.69) yields a symmetric matrix (16,16) as seen below which is
half filled.(The columns that follow after user 2 are omitted for readability
purposes).

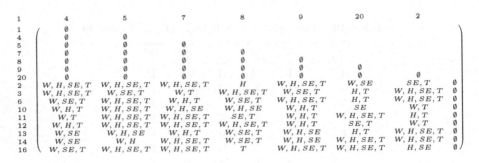

The result of applying of steps e through f and knowing that 4 attributes
are more than enough is displayed in tables 6.41 and 6.42. Deterministic Rules
extracted:

1. 5 Rules on Success:

   (a) If (T=M and SE=M) then User= SU
   (b) If (W=S and H=M) then User=SU
   (c) If (W=S and SE=S) then User=SU

Table 6.41. Core of the set of final data (Step e- of SSP)

Users	W	H	SE	T	E
7	S	M	S	M	F
20	S	S	M	S	F
9	M	S	L	M	F
1	L	X	X	X	F
4	L	X	X	X	F
5	L	X	X	X	F
8	L	X	X	X	F
2	S	S	S	M	SU
3	X	X	X	L	SU
11	X	X	X	L	SU
12	X	X	X	L	SU
6	X	X	X	L	SU
10	M	S	M	M	SU
13	M	M	L	VL	SU
16	S	M	S	S	SU
14	VL	X	X	X	SU

Table 6.42. Set of reductset (Step f- of SSP)

Users	W	H	SE	T	E
7	X	X	S	M	F
20	X	X	M	S	F
9	X	X	L	M	F
1,4,5,8	L	X	X	X	F
2	S	S	S	M	SU
16	S	M	S	S	SU
3,6,11,12	X	X	X	L	SU
10	X	X	M	M	SU
13	X	M	L	X	SU
14	VL	X	X	X	SU

  (d) If (T=L) then User=SU
  (e) If (T=VL) then User= SU
2. 4 Rules on Failure:
  (a) If (T=M and SE=L) then User= F
  (b) If (W=S, H=M and T=M) then User =F
  (c) If (SE=M and T=S) then User=F
  (d) If (W=L) then User= F

These 9 rules say that 4 parameters are needed to classify the 16 determin-istic examples or users out of the original 20. It seems that these deterministic rules are dependent of the attributes {W,H,SE,T}.

Reducts, reduct Sets(RS), and Minimum reduct Set(MRS)

In our example $\Omega=\{W,T,SE\}$ and $D=\{S\}$, and thus the power set of Ω is equal to:

$2^{\Omega} = \{\emptyset,\{W\},\{T\},\{SE\},\{WT\},\{WSE\},\{TSE\},\{WTSE\}\}$ K is said to be a minmimal RS or MRS iff it is a reductset RS with the minimum number of elements in it such that :

$$|K| \leq |R| \forall R \in 2^{\Omega}$$

Applying the concept of minimal reduct set to find deterministic rules
Thus the lower approximation of the set of being SU by applying (6.55) to table 6.39 is:

$I_*(SU)=\{2,3,6,10,11,12,13,14,16\}$

and the upper approximation of being SU by applying (6.56) is:

$I^*(SU)=\{2,3,6,10,11,12,13,14,16\}$.

Similarly in the concept of F, its lower approximation by applying (6.55) is:

$I_*(F)=\{1,4,5,7,8,9,20\}$

and its upper approximation by applying (6.56) is:

$I^*(F)=\{1,4,5,7,8,9,20\}$.

Table 6.39 has no boundary regions. By applying (6.62) and (6.63) to table 6.38, as we did above, yields the following:

$Ind(\{W\}) = \{\{2,7,11,16,20\}_S, \{3,6,9,10,12,13\}_M, \{1,4,5,8\}_L, \{14\}_{VL}\}$

$Ind(\{SE\}) = \{\{2,7,16\}_S, \{1,8,10,11,12,20\}_M, \{3,6,5,9,13\}_L, \{4,14\}_{VL}\}$

$Ind(\{T\}) = \{\{16,20\}_S, \{2,5,7,9,10\}_M, \{3,6,11,12\}_L, \{1,4,8,13,14\}_{VL}\}$

$Ind(\{H\}) = \{\{2,9,10,12,20\}_S, \{1,6,7,11,13,14,16\}_M, \{3,4,5,8\}_L\}$

$Ind\{W,H\} = \{\{2,20\}, \{7,11,16\}, \{9,10,12\}, \{6,13\}, \{3\}, \{4,5,8\}, \{1\}, \{14\}\}$

$Ind\{W,SE\} = \{\{2,7,16\}, \{11,20\}, \{10,12\}, \{3,6,9,13\}, \{1,8\}, \{5\}, \{4\}, \{14\}\}$

$Ind\{H,SE\} = \{\{2\}, \{10,12,20\}, \{7,16\}, \{1,11\}, \{6,13\}, \{8\}, \{9\}, \{3,5\}, \{4\}, \{14\}\}$

$Ind(\{W,T\} = \{\{20\}, \{2,16\}, \{7,11\}, \{9\}, \{3,6,10,12\}, \{13\}, \{5\}, \{1,4,8\}, \{14\}\}$

$Ind(\{T,SE\}) = \{\{20\}, \{2,16\}, \{5,9\}, \{7\}, \{10,11,12\}, \{3,6\}, \{1,8\}, \{13\}, \{4,14\}\}$

$Ind(\{H,T\} = \{\{20\}, \{12\}, \{16\}, \{2,9,10\}, \{7\}, \{5\}, \{6,11\}, \{3\}, \{1,13,14\}, \{4,8\}\}$

$Ind(\{W,T,SE\} = \{\{2\}, \{7,16\}, \{10,12\}, \{6,13\}, \{3\}, \{20\}, \{1\}, \{8\}, \{9\}, \{5\}, \{4\}, \{11\}, \{14\}\}$

$Ind\{W,H,SE\} = \{\{2\}, \{20\}, \{7,16\}, \{11\}, \{10,12\}, \{9\}, \{6,13\}, \{3\}, \{8\}, \{5\}, \{4\}, \{1\}, \{14\}\}$

$Ind(\{W,T,H\} = \{\{20\},\{2\},\{16\},\{7,11\},\{9\},\{10,12\},\{6\},\{3\},\{13\},$
$\{5\},\{1\},\{4,8\},\{14\}\}$
$Ind(\{H,SE,T\} = \{\{2\},\{20\},\{10\},\{12\},\{16\},\{7\},\{11\},\{1\},\{6\},\{13\},$
$\{8\},\{9\},\{5\},\{3\},\{4\},\{14\}\}$

Given B = $\{W,T\}$ and D = $\{E\}$ the positive boundary region is defined:
$POS\{W,T\}(E) = \bigcup_{x\in E}\{W,T\}_*(X) = \bigcup\{\{13,14\},\{5,9,20\}\}$
$= \{5,9,13,14,20\}$
Applying (6.60) or (6.61) to B and E yields:
$\gamma(\{W,T\},D) = |\{5,9,13,14,20\}|/16 = 5/16.$

Given B = $\{W,SE\}$ and D = $\{E\}$ the positive boundary region is defined:
$POS\{W,SE\}(E) = \bigcup_{x\in E}\{W,SE\}_*(X) = \bigcup\{\{14\},\{4,5\}\} = \{4,5,14\}$
Applying (6.60) or (6.61) to B and E yields:
$\gamma(\{W,SE\},D) = |\{4,5,14\}|/16 = 3/16.$

Given B = $\{SE,T\}$ and D = $\{E\}$ the positive boundary region is defined:
$POS\{SE,T\}(E) = \bigcup_{x\in E}\{SE,T\}_*(X) = \bigcup\{\{13\},\{7,20\}\} = \{7,13,20\}$
Applying (6.60) or (6.61) to B and E yields:
$\gamma(\{SE,T\},D) = |\{7,13,20\}|/8 = 3/16.$
Given B = $\{W,H\}$ and D = $\{E\}$ the positive boundary region is defined:
$POS\{W,H\}(E) = \bigcup_{x\in E}\{W,H\}_*(X) = \bigcup\{\{1\},\{3,14\}\} = \{1,3,14\}$
Applying (6.60) or (6.61) to B and E yields:
$\gamma(\{W,H\},D) = |\{1,3,14\}|/8 = 3/20.$

Given B = $\{T,H\}$ and D = $\{E\}$ the positive boundary region is defined:
$POS\{H,T\}(E) = \bigcup_{x\in E}\{H,T\}_*(X) = \bigcup\{\{5,7,20\},\{3,12,16\}\} =$
$\{3,5,7,12,16,20\}$
Applying (6.60) or (6.61) to B and E yields:
$\gamma(\{T,H\},D) = |\{3,5,7,12,16,20\}|/20 = 6/20.$

Given B = $\{H,SE\}$ and D = $\{E\}$ the positive boundary region is defined:
$POS\{H,SE\}(E) = \bigcup_{x\in E}\{SE,H\}_*(X) = \bigcup\{\{4,8,9\},\{2,14\}\}$
$= \{2,4,8,9,14\}$
Applying (6.60) or (6.61) to B and E yields:
$\gamma(\{SE,H\},D) = |\{2,4,8,9,14\}|/16 = 5/16.$

set of B's: B = $\{W,SE\}$, B = $\{SE,T\}$, and B = $\{rmW,H\}$ have the lowest $\gamma(\{B\},D)$. That set has a $\gamma(\{B\},D)=3/16$. It constitutes the minimal reductset $K = \{\{W,SE\},\{SE,T\},\{W,H\}\}$.

Deterministic rules extracted from table 6.39 by applying steps a-f of SSP as shown in table 6.43 and table 6.44 yield:

1. 5 Rules on Success:

    (a) If (T=M and SE=M) then User= SU.
    (b) If (W=S and H=M) then User=SU.

Table 6.43. Applying steps a-e of SSP to table 6.40

Users	W	H	SE	T	E
7	X	X	X	M	F
20	X	X	M	X	F
9	X	X	L	X	F
1	L	X	X	X	F
4	L	X	X	X	F
5	L	X	X	X	F
8	L	X	X	X	F
2	X	S	X	X	SU
11	X	X	X	L	SU
16	X	M	X	X	SU
6	X	X	X	L	SU
10	M	X	X	X	SU
12	X	X	X	L	SU
13	X	X	X	VL	SU
3	X	X	X	L	SU
14	VL	X	X	X	SU

Table 6.44. Set of reductset (Step f- of SSP)

Users	W	H	SE	T	E
7	X	X	X	M	F
20	X	X	M	X	F
9	X	X	L	X	F
1,4,5,8	L	X	X	X	F
2	X	S	X	X	SU
16	X	M	X	X	SU
3,6,11,12	X	X	X	L	SU
10	M	X	X	X	SU
13	X	X	X	VL	SU
14	VL	X	X	X	SU

  (c) If (W=S and SE=S) then User=SU.
  (d) If (T=L) then User=SU.
  (e) If (T=VL) then User= SU.
2. 4 Rules on Failure:

  (a) If (T=M and SE=L) then User= F.
  (b) If (W=S, H=M and T=M) then User =F.
  (c) If (SE=M and T=S) then User=F.
  (d) If (W=L) then User= F.

The minimal reductset yielded exact result to the idea Discernibility Function. These 9 deterministic rules are dependent on the 4 attributes: {W, H, SE, T}.

Decision Rules: Deterministic and Non Deterministic
Combined with the 9 deterministic rules that were specified above, 1 other non deterministic rule stand out:
1 Contradictory Rule:
If (W=S), (H=S), and (SE=S) and (T=S) then User= SU or F.
Contradictory rules also called inconsistent or possible or non deterministic rules have the same conditions but different decisions, so the proper decision cannot be made by applying this kind of rules. Possible decision rules determine a set of possible decision, which can be made on the basis of given conditions. With every possible decision rule, we will associate a credibility factor of each possible decision suggested by the rule. A consistent rule is given a credibility factor of 1, and an inconsistent rule is given a credibility factor smaller than 1 but not equal to 0. The closer it is to one the more credible the rule is. By applying 6.71, we have in this case a credibility factor of inconsistent rules $3/4 > .5$ which makes more credible in the case of success and $1/4$ in the case of Failure which makes less credible (being closer equal to 0). In fact the above considerations give rise to the question whether "Success" or "Failure" depends on {Searches, Time, Web Pages, Time}. As the above analysis, this is not the case in our example since the inconsistent rules do not allow giving a definite answer. In order to deal with this kind of situations, partial dependency of attributes can be defined.

Dependency Ratio = Number of deterministic Rules/Total Number of rules = $9/10 = .9$

6.6 Information Theory to Interpret User's Web Behavior

Learning can refer to either acquiring new knowledge or enhancing or refining skills. Learning new knowledge includes acquisition of significant concepts, understanding of their meaning and relationships to each other and to the domain concerned. The new knowledge has to be assimilated and put in a mentally usable form before it can be learned. Thus knowledge acquisition is defined as learning new symbolic information combined with the ability to use that information effectively. Skill enhancement applies to cognitive skills such as learning a foreign language or in our case Web behavior navigation. Leaning has been classified as either supervised or unsupervised. The distinction depends on whether the leaning algorithm uses pattern class information. Supervised learning assumes the ability of a teacher or supervisor who classifies the training examples into classes, whereas unsupervised learning must identify the pattern-class information as a part of the learning process. In supervised learning algorithms we utilize the information on the class membership of each training instance. This information allows supervised learning algorithms to detect pattern misclassifications as a feedback to themselves. ID_3 is an example of supervised learning. ID_3 uses a tree representation for concepts (Quinlan [113]). To classify a set of instances, we start at the top of the tree. And answer the questions associated with the nodes in the tree until we reach a leaf node where the classification or decision is stored. ID_3 starts by choosing a random subset of the training instances. This subset is called the window. The procedure builds a decision tree that correctly classifies all instances in the window. The tree is then tested on the training instances outside the window. If all the instances are classified correctly, then the procedure halts. Otherwise, it adds some of the instances incorrectly classified to the window and repeats the process. This iterative strategy is empirically more efficient than considering all instances at once. In building a decision tree, ID_3 selects the feature which minimizes the entropy function and thus best discriminates among the training instances. The ID_3 Algorithm:

1. Select a random subset W from the training set.
2. Build a decision tree for the current window:

   (a) Select the best feature which minimizes the entropy function H:

$$H = \sum_{i=1}^{n} -p_i lop(p_i) \tag{6.72}$$

   Where p_i is the probability associated with the ith class. For a feature the entropy is calculated for each value. The sum of the entropy weighted by the probability of each value is the entropy for that feature.

   (b) Categorize training instances into subsets by this feature.
   (c) Repeat this process recursively until each subset contains instances of one kind or some statistical criterion is satisfied.

3. Scan the entire training set for exceptions to the decision tree.
4. If exceptions are found, insert some of them into W and repeat from step 2. The insertion may be done either by replacing some of the existing instances in the window or by augmenting it with new exceptions.

6.6.1 ID_3 Applied to Case 1: Searches, Hyperlinks, and Web Pages([98])

By applying the ID_3 algorithm and equation 6.72 to the example in table 6.18 yields 9 rules on Success and failure. Rules extracted (See figure 6.32):

1. 2 Contradictory rules:
   (a) If (W=S), (H=S), and (SE=S) then User= SU or F.
   (b) If (W=S), (H=M), and (SE=S) then User= SU or F.
2. 5 Rules on Success:
   (a) If (W=VL) then User= SU (similar to 1.b.)
   (b) If (W=M , H=S and SE=M) then User= SU
   (c) If (W=M and H=L) then User= SU (similar to 1a.)

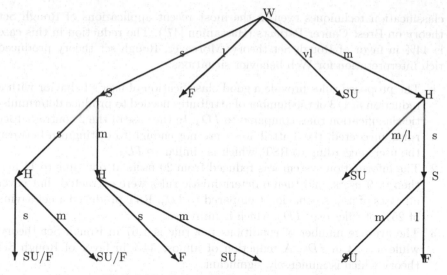

Fig. 6.32. Case 1: Rules extracted by ID_3

Table 6.45. Comparison of generated rules between RS Theory and ID_3

Algorithm	Det Rules			Non Det Rules
	Before Pruning	After Pruning	Average Conditions per rule	
ID_3	20	8	3	2
Rough Set	20	6	1.67	2

    (d) If (W=M and H=M) then User= SU
    (e) If (W=S , H=M, and SE=M) then User= SU
3. 3 Rules on Failure:
    (a) If (W=L) then User= F (similar to 2b.).
    (b) If (W=M , H=S, and SE=L) then User= F
    (c) If (W=S , H= S, and SE=M) then User= F

It seems that the number of rules extracted by ID_3 is more than the number of rules extracted by Rough set theory. The 3 parameters were not enough to separate these cases between success and failure as seen in Figure 6.32. Table 6.45 summarizes the number of generated rules between ID_3 and Rough Set theory. The ratio of Deterministic rules between Rough set theory and ID_3 after pruning is 6/8. The reduction is 25% which is a great number of rules if considered on the huge number of users and thus rules. The reduction of the number of parameters by Rough Set theory was 1/3 since the attribute SE becomes not needed to classify the users without loosing any accuracy in classification. Also the average number of conditions per rule is relatively low in Rough set theory compared to ID_3. The reduction is significantly high and has was never part of the comparison between Rough set theory and other classification techniques even in the most recent applications of Rough Set theory on Brest Cancer Patients (Hassanien [47]). The reduction in this case is 44% in favor of Rough Set theory. Moreover, Rough set theory produced rich interpretation for Web behavior situations:

1. The proposed rules provide a good classification of user's behavior with a reduction of 1/3 of the number of attributes needed to produce deterministic classification rules compared to ID_3. In the case of the 2 contradictory rules uncovered, the 3 attributes were not enough to distinguish between the users according to RST, which is similar to ID_3.
2. The information system was reduced from 20 users at one time to 11 and then to 9 users, and then 6 deterministic rules were extracted that cover all cases of user's behavior. Compared to ID_3, RST produced a net saving of 25% of rules over ID_3, which is immense.
3. The average number of conditions per rule is 1.67 in Rough Set theory, while it is 3 in ID_3. A reduction of almost 44% in favor of Rough Set theory which is immensely significant.

6.6.2 ID_3 Applied to Case 2: Searches, Time and Web Pages

By applying the ID_3 algorithm and equation 6.72 to the example in table 6.30 yields 9 rules on Success and failure.

Search: (SE)

$$
\begin{aligned}
&= (7/20)((-5/7)\log(5/7,2) - (2/7)\log(2/7,2)) + (6/20)(-(3/6) \\
&(\log(3/6,2) - (3/6)\log(3/6,2)) + (5/20)(-(2/5)\log(2/5,2) - \\
&(3/5)\log(3/5,2)) + (2/20)(-(1/2)\log(1/2,2) - (1/2)\log(1/2,2))) \\
&= 0.4799
\end{aligned}
$$

Web Pages(W):

$$
\begin{aligned}
&= (9/20)((-6/9)\log(6/9,2) - (3/9)\log(3/9,2)) + (6/20)(-(5/6)(\log(5/6,2) \\
&-(1/6)\log(1/6,2)) + (3/20)(-(3/3)\log(3/3,2)) + (2/20)(-(1/2)\log(1/2,2) \\
&-(1/2)\log(1/2,2))) = 0.4013
\end{aligned}
$$

Time(T):

$$
\begin{aligned}
&= (4/20)((-2/4)\log(2/4) - (2/4)\log(2/4)) + (5/20)(-(3/5)(\log(3/5) \\
&-(2/5)\log(2/5)) + (6/20)(-(5/6)\log(5/6) - (1/6)\log(1/6))) + (5/20) \\
&(-(3/5)\log(3/5) - (2/5)\log(2/5))) = 0.3432
\end{aligned}
$$

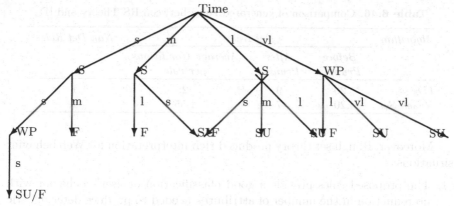

Fig. 6.33. Case 2: Rules extracted by ID_3

Under Time=Small, we calculate Search, Web Pages.
Small:
Search = $(4/4)((-2/4)\log(2/4) - (2/4)\log(2/4)) = 1$
WebPages = $(3/4)((-2/3)\log(2/3) - (1/3)\log(1/3)) = 0.6887$

The smallest is W.
Rules extracted (See figure 6.33):

1. 1 Contradictory rule:

   (a) If (W=S), (H=S), and (SE=S) then User= SU or F.

2. 5 Rules on Success:

   (a) If (T=VL and W=M) then User= SU
   (b) If (T=VL and W=VL) then User= SU
   (c) If (T=M and SE=S) then User= SU
   (d) If (T=L and SE=L) then User= SU
   (e) If (T=L and SE=M) then User= SU

3. 4 Rules on Failure:

   (a) If (T=S and SE=M) then User= F (similar to 2b.)
   (b) If (T=M and SE=L) then User= F (similar to 2a.)
   (c) If (T=L and SE=S) then User= F
   (d) If (T=VL and W=L) then User= F

It seems that the number of rules extracted by ID_3 is more than the number of rules extracted by Rough set theory. The 3 parameters were not enough to separate these cases between success and failure as seen in Figure 6.33. Table 6.46 summarizes the number of generated rules between ID_3 and Rough Set theory. The ratio of deterministic rules between Rough set theory and ID_3 after pruning is 7/9. The reduction is almost 22% which is a high percentage especially if the study were to involve a huge number of users and thus rules. That makes Rough Set theory a de facto technique for large experiments.

Table 6.46. Comparison of generated rules between RS Theory and ID$_3$

Algorithm	Det Rules			Non Det Rules
	Before Pruning	After Pruning	Average Conditions per rule	
ID$_3$	20	9	2	1
Rough Set	20	7	1.55	1

Moreover, Rough set theory produced rich interpretation for Web behavior situations:

1. The proposed rules provide a good classification of user's behavior with no reduction in the number of attributes needed to produce deterministic classification rules compared to ID_3. In the case of the contradictory rule uncovered, the 3 attributes were not enough to distinguish between the users according to RST and also to ID_3.
2. The information system was reduced from 20 users at one time to 12 and then to 8 users, and then 7 deterministic rules were extracted that cover all cases of user's behavior. Compared to ID_3, RST produced a net saving of 22% of rules over ID_3, which is immense.
3. The average number of conditions per rule is 1,55 in Rough Set theory, while it is 2 in ID_3. A reduction of almost 22% in favor of Rough Set theory which is very significant.

6.6.3 ID_3 Applied to Case 3: Web Pages, Hyperlink, Searches and Time

By applying the ID_3 algorithm and equation 6.72 to the example in table 6.38 yields 9 rules on success and failure.

Search: (S)

$$= (7/20)((-5/7)\log(5/7) - (2/7)\log(2/7)) + (6/20)(-(3/6)$$
$$(\log(3/6) - (3/6)\log(3/6)) + (5/20)(-(2/5)\log(2/5)$$
$$-(3/5)\log(3/5)) + (2/20)(-(1/2)\log(1/2) - (1/2)\log(1/2)))$$
$$= 0.47991$$

Hyperlinks:

$$= (9/20)((-6/9)\log(6/9) - (3/9)\log(3/9)) + (7/20)(-(5/7)(\log(5/7)$$
$$-(2/7)\log(2/7)) + (4/20)(-(1/4)\log(1/4) - (3/4)\log(3/4)) = .46228$$

Web Pages:

$$= (9/20)((-6/9)\log(6/9) - (3/9)\log(3/9)) + (6/20)(-(5/6)(\log(5/6)$$
$$-(1/6)\log(1/6)) + (3/20)(-(3/3)\log(3/3)) + (2/20)(-(1/2)\log(1/2)$$
$$-(1/2)\log(1/2))) = .40128$$

Time:

$$= (6/20)((-4/6)\log(4/6) - (2/6)\log(2/6)) + (5/20)(-(3/5)(\log(3/5)$$
$$-(2/5)\log(2/5)) + (4/20)(-(4/4)\log(4/4)) + (5/20)(-(3/5)\log(3/5)$$
$$-(2/5)\log(2/5))) = 0.367402$$

Under Small and Medium and Very Large Time,

we need to calculate Search, Web Pages, and hyperlinks: Small:

$Search = (5/6)((-4/5)\log(4/5) - (1/5)\log(1/5)) = 0.60161$
$Hyperlinks = (5/6)(-(2/5)\log(2/5) - (3/5)\log(3/5)) = 0.80913$
$WebPages = (-(4/6)\log(4/6) - (2/6)\log(2/6)) = 0.9183$

The smallest is Search: Medium:
$Search = (2/5)(-(1/2)\log(1/2) - (1/2)\log(1/2)) = .4$
$Hyperlinks = (3/5)(-(1/3)\log(1/3) - (2/3)\log(2/3)) = 0.5509775$
$WebPages = (2/5)(-(1/2)\log(1/2) - (1/2)\log(1/2)) - (2/5)(-(1/2)$
$\log(1/2) - 1/2)\log(1/2)) = .8$

The smallest is Search: VLT:
$Search = (2/5)(-(1/2)\log(1/2) - (1/2)\log(1/2)) = .5$
$Hyperlinks = (3/5)(-(1/3)\log(1/3) - (2/3)\log(2/3)) = 0.6887219$
$WebPages = 0$

The smallest is Web Pages

Under Small Search we calculate Web Pages, and hyperlinks:
Small:
$Hyperlinks = (4/5)(-(3/4)\log(3/4, 2) - (1/4)\log(1/4, 2)) = 0.649022$
$WebPages = (-(4/5)\log(4/5, 2) - (1/5)\log(1/5, 2)) = 0.7219281$

Rules extracted (See figure 6.34):

(a) 1 Contradictory rule:
    a) If (T=S) and (W=S) and (H=S), and (SE=S) then User= SU/F.
1. 6 Rules on Success:
    (a) If (T=S and SE=S and H=M) then User= SU
    (b) If (T=M and SE=M) then User= SU (Similar to 1a.)
    (c) If (T=VL and W=M) then User= SU
    (d) If (T=VL and W=VL) then User= SU
    (e) If (T=M and SE=S and H=S) then User =SU
    (f) If (T=L) then User =S
2. 4 Rules on Failure:
    (a) If (T=VL and W=L) then User= F
    (b) If (T=M and SE=S and H=M) then User =F

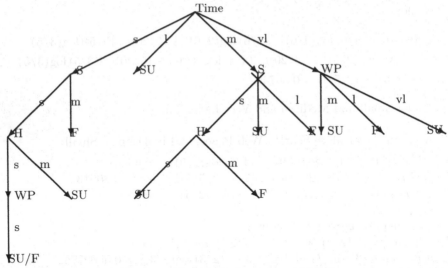

Fig. 6.34. Case 3: Rules extracted by ID_3

Table 6.47. Comparison of generated rules between RS Theory and ID_3

Algorithm	Det Rules			Non Det Rules
	Before Pruning	After Pruning	Average Conditions per rule	
ID_3	20	10	2.2	1
Rough Set	20	9	1.77	1

(c) If (T=S and SE=M) then User= F(Similar to 2c.)
(d) If (T=M and SE=L) then User= F(similar to 2a.)

Table 6.47 summarizes the number of generated rules between ID_3 and Rough Set theory. The ratio of deterministic rules between Rough set theory and ID_3 after pruning is 9/10. The reduction is 10% which is still significant if the study were to involve a huge number of users and thus rules. That makes Rough Set theory a de facto technique for large experiments.

Moreover, Rough set theory produced rich interpretation for Web behavior situations:

1. The proposed rules provide a good classification of user's behavior with no reduction in the number of attributes needed to produce deterministic classification rules compared to ID_3. In the case of the contradictory rule uncovered, the 4 attributes were not enough to distinguish between the users according to RST and also to ID_3.

2. The information system was reduced from 20 users at one time to 18 and then to 16 users, and then 9 deterministic rules were extracted that cover all cases of user's behavior. Compared to ID_3, RST produced a net saving of 10% of rules over ID_3, which is significant.
3. The average number of conditions per rule is 1.77 in Rough Set theory, while it is 2.2 in ID_3. A reduction of almost 19.5% in favor of Rough Set theory which is a slight reduction.

6.6.4 Final Observation on Rough Set Theory and Information Theory Final to Interpret User's Web Behavior

This application of the Rough set methodology shows the suitability of the approach for the analysis of user's Web information system's behavior. Rough set theory was never applied on user's behavior and the latter was analyzed very little (Meghabghab [98]) considering the amount of emphasis on understanding user's Web behavior ([73]). Moreover, Rough set theory produced rich interpretation for Web behavior in three different conditional situations as seen in table 6.48:

1. The proposed rules provide a good classification of user's behavior with a reduction in one case of 1/3 of the number of attributes, in case 1, needed to produce deterministic classification rules compared to ID_3. In the case of the contradictory rules uncovered, in all 3 cases considered, the number of attributes was not enough to distinguish between the users according to RST and also to ID_3.
2. The information system was reduced from 20 users at one time to 12 and then to 8 users, and the number of deterministic rules that were extracted varied from 6, in case 1 which is the lowest, to 9, in case 3 which is the highest, in the case of RST. Compared to ID_3, RST produced a net saving of 25% of rules over ID_3, in case 1 which is the best case, to 10% in case which is the worst case, and which is still very significant.
3. The average number of conditions per rule was always lower in RST than in ID_3. The reduction reached a 44%, in case 1 which is the best case, to 19.5% in case 3 which is the worst case, which is still slightly significant.

Table 6.48. Summary of Comparison between RS theory and ID_3

Ω	Rough Set			ID_3		
	Ω	Det Rules	ACPR	Ω	Det Rules	ACPR
{W,H,SE}	{W,H}	6	1.67	{W,H,SE}	8	3
{W,T,SE}	{W,T,SE}	7	1.55	{W,T,SE}	9	2
{W,H,T,SE}	{W,H,T,SE}	9	1.77	{W,H,T,SE}	10	2.2

6.6.5 General Summary of Rough Set Theory

The Rough set (RS) theory proposed by Pawlak ([106]) is concerned with qualitative data analysis. The data set being analyzed consists of objects (also examples, cases) which may represent states, Web users, stocks, etc., described by values of the attributes, representing features, parameters, etc. of the objects. The set of attributes is divided into two disjoint subsets, called condition and decision attributes. The important distinction between these two sets is that the condition attributes express some descriptive information about the objects, while the decision attributes express some decisions or conclusions made about the objects. The set of objects described by attributes and represented in a table form is called a decision table. One of the fundamental notions of the RS theory is the indiscernibility relation. Indiscernibility identifies groups of objects which are indiscernible from one another when only some of the attributes are taken into account. The relation is based on the fact that the values of attributes are the sole source of knowledge about the objects. Indiscernibility is an equivalence relation, so it defines a partition of objects into disjoint subsets. The main concern of the RS theory is to examine different partitions induced by sets of condition and decision attributes and the relationship between these two partitions. Two particular partitions of the objects are of special interest. One of them is the partition induced by the set of all the decision attributes. This partition defines classes of objects - namely the sets of objects described by the same values of the decision attributes. Because often there is only one decision attribute in the table, the classes are sets of objects featuring the same value of the decision attribute. The other partition of interest is induced by the set of all the condition attributes. The elements of this partition, called atoms, contain objects which are indiscernible from one another with regard to all of the condition attributes. The name 'atom' is intended to stress the fact that it is possibly the smallest and unsplittable 'granule' of information. Any two objects belonging to the same atom are consequently described by the same set of values and should be treated in exactly the same way in further analysis. Defining the atoms and classes allows for the next step of the RS analysis in which the equivalence sets induced by the condition attributes (atoms) and the decision attributes (classes) are matched to check if the classes can be expressed by the atoms. Any subset of objects (called a concept) is definable by some attributes if it may be represented as a union of elements of the partition generated by this set of attributes. In that sense the concept is built of atoms. A very interesting situation arises in case of inconsistency, that is when a concept is not definable by a subset of attributes, and thus it cannot be constructed from the atoms, because there are atoms in the decision table which contain objects both within the concept's definition and outside it. Because the atoms cannot be split any further, the RS theory introduces a notion of approximation and instead of representing the concepts it represents their approximations.

Lower approximation of a concept is a set of objects belonging to atoms which are completely included in the concept's definition. If an atom contains objects of which only some belong to the concept's definition, such an atom constitutes the upper approximation of the concept. Final notion of interest to the RS analysis is a boundary region which is defined as the difference between the upper and lower approximations. In the case when inconsistency does not occur and the concepts are definable in terms of atoms, then both the lower and upper approximations of the concept are simply equal to each other. This implies that a boundary region is empty. The RS theory introduces a measure of inconsistency, called the quality of approximation, which is defined as the ratio of the number of all objects belonging to lower approximations of all classes to the number of all objects in the decision table. Maximum value of this measure, being 1.0, indicates that all the classes may be fully defined using the condition attributes. If the quality of approximation is satisfactory, then it may be interesting to see if there are some subsets of condition attributes which are sufficient to generate the same quality of approximation as a whole set. This notion leads directly to the idea of attributes' reduction.

Reducts and Their Computation

A reduct is defined as a subset of attributes which ensures the same quality of approximation as the original set. In general, it is possible that there is more than one reductfor a given decision table. In that case the set called the core of attributes is defined as the intersection of all reducts. Removal of any single attribute included in the core always leads to the drop of the quality of approximation. Generating reducts is a computationally complex task and normally two classes of reduct generating algorithms are used:

- Exact algorithms: designed for generating all reducts from a decision table. Their main disadvantage is the computational complexity that may grow non-polynomially with the number of attributes in the table.
- Approximate algorithms: designed for generating single reducts, with the aim of reducing the computing time. This allows for avoiding the non-polynomial computational load but it produces approximate solutions which are not true reducts.

Decision Rules

A very important part of the RS analysis concerns expressing the dependencies between the values of the condition and the decision attributes. The main task is to find these dependencies and represent them in an easily interpretable manner. One possible way of doing this is through the construction of decision rules. A decision rule is expressed as a logical statement consisting of the condition and decision parts. The interpretation of the rule is as follows: "IF the condition attributes assume values indicated in the condition part of the rule THEN the value of decision attribute(s) is one of those indicated by the decision part". Decision rule may be categorized as consistent if all objects matching its condition part have the value which is indicated by the decision part of the rule, otherwise a rule is inconsistent; and it may be categorized as exact if the decision part of the rule contains only one elementary condition, otherwise a rule is approximate. Approximate rule is a result of an approximate description of concepts in terms of atoms. It implies that using the available knowledge, it is not possible to decide whether some objects (from the boundary region) belong to a given decision class or not. Each decision rule is characterized by its strength, which is defined as the number of objects satisfying the condition part of the rule (or, in other words, which are covered by a rule) and belonging to a given decision class. In the case of approximate rules, the strength is calculated separately for each possible decision class. Stronger rules are often more general, i.e. their condition parts are shorter and less specialized, and they are usually more interesting from an analyst's point of view. Procedures for generating decision rules from decision tables are mostly based on induction. The existing inductive algorithms use one of the following strategies to produce the set of decision rules:

1. generation of a minimal set of rules covering all objects,
2. generation of an exhaustive set consisting of all possible rules,
3. generation of a set of "strong" decision rules, consistent or not, each covering relatively many objects, but not necessarily all of them.

An important feature of the RS methodology is that the rules for a given decision class are actually generated not from the concept definition but from its approximations. Lower approximation serves first to produce exact rules for the class, while the boundary regions are subsequently used to generate approximate rules that cover the inconsistencies.

Classification of New Objects

Decision rules derived from a decision table can be used for recommendations regarding the classes of new objects (classification). Specifically, the classification of a new object can be supported by matching its description to one of the Decision Rules. This matching may lead to one of three situations ([121]):

1. the new object matches exact rules indicating one decision class,
2. the new object matches exact or approximate rules indicating different decision classes,
3. the new object does not match any of the rules.

In case of situation (1) the recommendation is univocal. In the case of ambiguous matching (situation (2)), the final classification is made on a basis of rule strengths. For each class which is indicated by any of the rules, a coefficient called class support defined as the total strength of all the rules supporting this class is computed. The decision class featuring the highest value of the class support is assigned as a class for the new object. In the case of situation (3), an object may remain unclassified. However, some other method (such as the default class classifier) for predicting an object's class may be used in this case. The default classifier assigns an object to a selected class, so called default class (each object is assigned the same class). Usually the most frequently occurring class in a decision table is selected to be a default class, in which case a classifier is referred to as majority class classifier. In summary, the process of classifying objects may actually consist of two phases, the first of them being classification using the rules, while the second involves handling the still unclassified objects. These objects may be assigned class membership according to, for example, a majority class classifier.

References

1. Search Engine Watch: Up-to-date information on leading search engines. Located at http://www.searchenginewatch.com/.
2. D. Agosto, Bounded rationality and satisficing in young people's web-based decision making, JASIST 53(1) (2002) 16-27.
3. R. Albert, H. Jeong, A.L. Barabasi, Diameter of the World Wide Web, Nature 401 (1999) 130-131.
4. R. Baeza-Yates, B. Ribeiro-Neto, Modern Information Retrieval, Addison Wesley, 1999.
5. G.A. Banini, R.A.Bearman, Application of fuzzy cognitive maps to factors affecting slurry rheology. International Journal of Mineral Processing 52(4) (1998) 233-244.
6. K. Bharat, M.R. Henzinger, Improved algorithms for topic distillation in a hyper-linked environment, Proceedings of the ACM Conference on Research and Development in Information Retrieval, 1998.
7. K. Bharat, A. Broder, A technique for measuring the relative size and overlap of public Web search engines, Proceedings of the 7th International world wide web conference, Brisbane, Australia, 1998, pp. 379-388.
8. R. Biswas, An application of Yager's Bag Theory in multi-criteria based decision making problems, International Journal of Intelligent Systems 14 (1998) 1231-1238.
9. V.D. Blondel, P. Van Dooren, A measure of similarity between graph vertices, In Proc. of the ICALP 2003 Conference, Lecture Notes in Computer Science, Volume 2719, J.C.M. Baeten et al. Eds., Spinger Verlag, pp.739-550.
10. D.P. Bosc, D. Rocacher, About the difference operation on fuzzy bags, IPMM 02, Annecy, France, 2002, pp. 1541-1546.
11. S. Brin, Extracting patterns and relations from the World Wide Web, Proceedings of WebDB'98, Valencia, Spain, 1998.
12. S. Brin, L. Page, The anatomy of a large scale hypertextual web search engine, Proceedings of the 7th World Wide Web Conference, Brisbane, Australia, 1998.
13. R. Botafogo, E. Rivlin, B. Shneiderman, Structural analysis of hypertexts: Identifying hierarchies and useful metrics, ACM Transactions on Information Systems, 10 (1992) 142-180.
14. H. J. Brockman, Digerati: Encounters With the Cyber Elite, Hardwired, 1996.

15. R. Broder, F. Kumar, P. Maghoul, R. Raghavan, R. Stata, Graph structure in the Web, In Proc. of the 9th International World Wide Web Conference, 2000, pp. 247-256.

16. Bruce, What do we know about academic users, In Proceedings of the Mid-Year Conference of the American Society for Information Science, Pasadena, CA, 1999.

17. O. Case, Looking for information: a survey of research on information seeking, needs, and behaviour, Amsterdam: Academic Press, 2002.

18. S. Chakrabarti, B. Dom, D. Gibson, J. Kleinberg, P. Raghavan, S. Rajagopalan, Automatic resource compilation by analyzing hyperlink structure and associated text, Proceedings of the 7th International World Wide Web Conference, Brisbane, Australia, 1998.

19. S. Chakrabarti, B. Dom, D. Gibson, S.R. Kumar, P. Raghavan, S. Rajagopalan, A. Tomkins, Experiments in topic distillation, ACM SIGIR Workshop on Hypertext Information Retrieval on the Web, 1998.

20. D. Canter, R. Rivers, G. Storrs, Characterizing User Navigation through Complex Data Structures, Behavior and Information Technology 4 (2) (1985).

21. C.H. Chang, and C.C. Hsu, Integrating query expansion and conceptual relevance feedback for personalized web information retrieval. 7th International World Wide Conference, Brisbane Australia, 14-18 April, 1998.

22. C.W. Choo, B. Detlor, D. Turnbull, Information Seeking on the Web: An Integrated Model of Browsing and Searching, Proceedings of the Annual Conference of the American Society for Information Science, Washington D.C., 1999, pp.127-135.

23. K. Chung, Spectral Graph Theory, American Mathematical Society, 1997.

24. F.R. Connan, F.R, Interrogation Flexible de bases de donnees multi-medias, These D'etat, University of Rennes I, (1999).

25. T.M. Cover, Geometrical and statistical properties of systems of linear inequalities with applications in pattern recognition, IEEE Transactions on Electronic Computes vol 14 (1965) 326-334.

26. P. Craiger, M.D. Coovert, "Modeling dynamic social and psychological processes with fuzzy cognitive maps," IEEE. (1994).

27. I. De Sola Pool, Technologies without Boundaries, Harvard University Press, 1990.

28. L. DiLascio, E. Fischetti, V. Loia, Fuzzy Management of user actions during hypermedia navigation, International Journal of approximate reasoning 18 (1998) 271-303.

29. J.A. Dickerson, B. Kosko, Virtual Worlds as Fuzzy Cognitive Maps, Presence 3(2) (1994) 173-189.

30. J.A. Dickerson, B. Kosko, Virtual Worlds as Fuzzy Dynamical Systems, In Technology for Multimedia, B. Sheu and M. Ismail(Editors), IEEE Press, New York, 1998, pp. 567-603.

31. D.M. Edwards, L. Hardman, Lost in hyperspace: cognitive mapping and navigation in a hypertext environment. In R. McAleese (Ed.) Hypertext: Theory and Practice, Oxford: Intellect, 1989.

32. L. Egghe, R. Rousseau, Introduction to infometrics, Elsevier, 1990.

33. D. Ellis, Modeling the information-seeking patterns of academic researchers: a grounded theory approach, Library Quarterly 63(4) (1993) 469-486.

34. D. Engelbart, Augmenting human intellect: a conceptual framework, Menlo Park, CA: Stanford Research Institute, 1962.

35. B.R. Fernandez, A. Smith, A. 7, 267-272, ASME Press: New York, 1997.
36. R. Fidel et al., A visit to the information mall: Web searching behavior of high school students, JASIS 50(1) (1999) 24-38.
37. M.E. Frisse, Searching for information in a hypertext medical handbook, CACM, 31 (1997) 880-886.
38. E. Garfield, Citation analysis as a tool in journal evaluation, Science 178 (1972) 471-479.
39. R. Gedge, R. (1997). Navigating in hypermedia–interfaces and individual differences, http://www.coe.uh.edu/insite/elec_pub/HTML1997/re_gedge.htm.
40. A. Gento, A. Redondo, Rough sets maintenance in a production line, Expert Systems 20(5) (2003) 271-279.
41. G. Golub, C.F. Van Loan, Matrix computations. Baltimore, Maryland, John Hopkins University Press, 1989.
42. M. Gross, Imposed queries in the school library media center: A descriptive study, Library and Information Science Research 21 (1999) 501-521.
43. S. Grumbach, T.M. Milo, Towards tractable algebras for multi-sets, ACM principles of Database Systems, (1993).
44. M. Hagiwara, Extended fuzzy cognitive maps. FUZZ-IEEE1992, IEEE International Conference on Fuzzy Systems, Mar 8-12, San Diego, CA, 1992, pp. 795-801.
45. N. Hammond, Learning with hypertext, In C. McKnight, A. Dillon, and J. Richardson (Eds), Hypertext, a psychological Perspective. London: Ellis Horwood,
46. C-G Han, P.M. Pardalor, Y. Ye, Computational aspect of an interior point algorithm for quadratic programming problems with box constraints, In Large Scale Numerical Optimization, edited by T.F. Coleman and Y. Li, Philadelphia, USA: SIAM publications, 1990, pp. 92-112.
47. A. Hassanien, Rough Set Approach for attribute reduction and rule generation: A case of Patients with suspected breast cancer, JASIST 55(11) (2004) 954-962.
48. B. Hassibi, D.G. Stork, Second order derivatives for nework pruning: Optimal brain surgeon, In Advances in Neural Information Processing Systems, 5, (NIPS), edited by S.J. hanson and J. Cowan, 1993, pp. 164-171.
49. T.H. Haveliwala, Topic-Sensitive PageRank, In D. Lassner, D. De Roure, and A. Iyengar, editors, Proc. 11th International World Wide Web Conference, ACM Press, 2002, pp. 517-526.
50. D. Hawking, N. Craswell, P. Bailey, K. Griffiths, Measuring search engine quality, Information Retrieval 4(1) (2001) 33-59.
51. S. Haykin, Neural networks: A comprehensive foundation, Prentice Hall, Second Edition. New York, 1999.
52. D.O. Hebb, "The Organization of Behavior: A Neuropsychological Theory" Wiley, New York, NY, 1949.
53. M.R. Henzinger, A. Heydon M. Mitzenmacher, M. Najork, On Near-Uniform URL Sampling, Ninth International WWW Conference, Amsterdam, May 15-19, 2000.
54. M. Higgins, Meta-information and time: Factors in human decision making, JASIST 50 (1999) 132-139.
55. W. Hersh, Information retrieval at the millennium, In Proceedings of AMIA, 1998, Lake Buena Vista, FL, 1998, pp.38-45.
56. M. Higgins, Meta-information and time: Factors in human decision making. JASIST 50 (1999) 132-139.

57. J. Hong, J.A. Dickerson, Fuzzy Agents for Vehicle Behavior Modeling in Driving Simulations, Smart Engineering Systems: Neural Networks, Fuzzy Logic, Data Mining, and Evolutionary Programming, ANNIE '97, St. Louis, edited by C.H. Dagli, M. Akay, O. Ersoy, 1997

58. R.D. Hurrion, Using a neural network to enhance the decision making quality of a visual interactive simulation model, Journal of the Operations Research Society 43(4) (1992) 333-341.

59. P. Ingwersen, P. K. Jarvelin (forthcoming), The turn: integration of information seeking and retrieval in context, Berlin: Springer/Kluwer.

60. J. Janes, The emergence of the web as a reference resource and tool: initial empirical results, In Proceedings of the Mid-Year Conference of ASIS. Pasadena, CA, 1999.

61. C. Johnson, Choosing people: the role of social capital in information seeking behaviour, Information Research 10(1) (2004) 1-17.

62. Y. Kafai, M.J. Bates, Internet web-searching instruction in the elementary classroom: building a foundation for information literacy, School Library Media Quarterly 25 (1997) 103-111.

63. M.M. Kessler, Bibliographic coupling between scientific papers, American Documentation 14, (1963) 10-25.

64. J. Kleinberg, Authoritative sources in a hyper-linked environment, Proceedings of the ACM-SIAM Symposium on Discrete Algorithms, 1998, pp. 668-677.

65. J. Kleinberg, Personal email to the first author, 2000.

66. T. Kopetzky, J. Kepler, Visual Preview of the link traversal on the www, Proceedings of the 8th International WWW confernce, Toronto, Canada, 1999.

67. B. Kosko, Fuzzy Cognitive Maps, International Journal Man-Machine Studies 24 (1986) 65-75.

68. B. Kosko, Hidden Patterns in Combined and Adaptive Knowledge Networks, International Journal of Approximate Reasoning 2 (1988) 337-393.

69. B. Kosko, Fuzzy Engineering. Prentice Hall, 1997.

70. R. Kumar, P. Raghavan, S. Rajagopalan, A. Tomkins, Trawling the web for emerging cyber communities, In Proceedings of The 8th International World Wide Web Conference, 1999.

71. R. Larson, Bibliometrics of the World Wide Web: An exploratory analysis of the intellectual structure of cyberspace, Annual Meeting of the American Society for Information Science, 1996.

72. A. Law, D. Kelton, Simulation Modeling and Analysis, McGraw Hill, New York, NY, 1991.

73. A.W. Lazonder, J.A. Biemans, G.J.H. Wopereis, Differences between novice and experienced users in searching information on the World Wide Web, JASIST 51 (6) (2000) 576-581.

74. G.J. Leckie, K.E. Pettigrew, C. Sylvain, Modeling the information seeking of professionals: a general model derived from research on engineers, health care professionals, and lawyers, Library Quarterly 66(2) (1996) 161-193.

75. K.S. Lee, S Kim, M. Sakawa, On-line fault diagnosis by using fuzzy cognitive maps. IEICE Transactions on Fundamentals of Electronics, Communications and Computer Sciences, E79-A 6 (1996) 921-922.

76. P. Lyman, H.R. Varian, How Much Information, Retrieved from http://www.sims.berkeley.edu/how-much-info on June 17, 2002.

77. K. Martzoukou, A review of Web information seeking research: considerations of method and foci of interest, Information Research 10(2) (2005) 1-27.

78. O. McBryan, GENVL and WWW: Tools for taming the Web, Proceedings of the 1st International WWW Conference, 1994.

79. J. McLuhan, Understanding Media: The Extensions of Man, Cambridge: MIT Press, 1995.

80. D.A. Menasce, Capacity Planning of Web Services: metrics, models, and methods, Prentice Hall, 2002.

81. G. Meghabghab, Stochastic Simulations of Rejected World Wide Web Pages, Proceedings of the IEEE 8th International Symposium on Modeling, Analysis, and Simulation of Computer and Telecommunications Systems (MASCOTS 2000), Aug 29-Sept 1, San Francisco, CA, pp. 483-491.

82. G. Meghabghab, D. Meghabghab, INN: An Intelligent Negotiating Neural Network for Information Systems: A Design Model, Inf. Process. Manage. 30(5) (1994) 663-686.

83. D. Meghabghab, G. Meghabghab, Information retrieval in cyberspace, In Proceedings of ASIS Mid-Year Meeting, 18-22 May, 1996, San Diego, CA, pp. 224-237.

84. G. Meghabghab, Mining User's Web Searching Skills Through Fuzzy Cognitive State Map, Joint 9th IFSA World Congress and 20th NAFIPS International Conference,Vancouver, CA, 2001, pp. 429-436.

85. G. Meghabghab, Fuzzy cognitive state map vs markovian modeling of user's web behavior, Systems, Man, and Cybernetics, IEEE International Conference on Volume 2, 7-10 Oct., 2001 pp. 1167-1172.

86. G. Meghabghab, D. Windham, Mining user's Behavior using crisp and fuzzy bags: A new perspective and new findings, NAFIPS-FLINT 2002 International Conference, New Orleans, LA, 2002, pp. 75-80.

87. G. Meghabghab, A. Kandel, Stochastic simulations of web search engines: RBF versus second-order regression models, Information Sciences 159 (1-2) (2004) 1-28.

88. G. Meghabghab, D. Bilal, Application of Information Theory to Question Answering Systems: Towards An Optimal Strategy, JASIS 42(6) (1991) 457-462.

89. G. Meghabghab, Iterative radial basis functions neural networks as metamodels of stochastic simulations of the quality of search engines in the World Wide Web, Information Processing and Management 37(4) (2001) 571-591.

90. G. Meghabghab, Discovering authorities and hubs in different topological web graph structures, Information Processing and Management 38 (2002) 111-140.

91. G. Meghabghab, Google's Web Page Ranking Applied to Different Topological Web Graph Structures, JASIST 52(9) (2001) 736-747.

92. G. Meghabghab, A Multi-criteria fuzzy bag for mining web user's behavior, Fuzzy Information Processing Theories and Applications, Beijing, China, March 1-4, 2003, Vol.2, pp 821-826.

93. G. Meghabghab, The Difference between 2 Multidimensional Fuzzy Bags: A New Perspective on Comparing Successful and Unsuccessful User's Web Behavior, LCNS 2003 pp. 103-110.

94. G. Meghabghab, (invited Paper), Losing Hubs and Authorities in Bow Tie graphs (BTG): Blondel's et al. to the rescue of the similarity of BTG, In Proc. of the ICDAT 2004 Conference, March 18-19, 2004, Taipei, Taiwan.

95. G. Meghabghab, Similarity between Bow Tie Graphs (BTG): A Centrality Score Analysis of BTG, Software Engineering Research and Practice, 2004, pp. 29-35.

96. G. Meghabghab, A fuzzy similarity measure based on the centrality scores of fuzzy terms, Proceeding NAFIPS IEEE Annual Meeting of the, Vol 2, 27-30 June 2004, pp. 740-744.

97. G. Meghabghab, Mining User's Web Searching Skills: Fuzzy Cognitive State Map vs. Markovian Modeling (Invited Paper), International Journal of Computational Cognition, Vol 1, No 3, September 2003, pp. 51-92,

98. G. Meghabghab, A Rough Set Interpretation of User's Web Behavior: A Comparison with Information Theoretic Measure, FLAIRS Conference 2005, pp. 283-288.

99. G. Meghabghab, D. Meghabghab, Multiversion Information Retrieval: Performance Evaluation of Neural Networks vs. Dempster-Shafer Model, 3rd Golden west international conference – 1994 Jun : Las Vegas; NV, Kluwer Academic; 1995, pp. 537-546

100. G. Meghabghab, J. Ji, Algorithmic enhancements to a backpropagation interior point learning rule, IEEE International Conference on Neural Networks, Volume 1, 3-6 June 1996, pp. 490-495.

101. G. Meghabghab, G. Nasr, An Interior-Point RBF Neural Network Algorithm on Four Benchmarks Problems, In Proceedings of the WORLD CONGRESS ON EXPERT SYSTEMS,1998, VOL 2 , Mexico City, Mexico, pp. 848-855.

102. G. Meghabghab, G. Nasr, Iterative RBF Neural Networks as Metamodels of Stochastic Simulations, International Conference on Intelligent Processing and Manufacturing of Materials, 1999, Vol. 2, Honolulu, HI, pp. 729-734.

103. S. Mukherjea, J. Foley, S. Hudson, Visualizing complex hypermedia networks through multiple hierarchical views, Proceedings of ACM SIGCHI Conference in Human Factors in Computing, Denver, CO, 1995, pp. 331-337.

104. R.H.. Myers, A.I. Khuri, G. Vining, Response Surface Alternatives to the Taguchi Robust Parameter Design Approach, American Statistician 46(2) (May 1992) 131-139.

105. M.R. Nelson, We Have the information you want, but getting tt will cost you: being held hostage by information overload, Crossroads 1(1) (1994) http://info.acm.org/crossroads/ xrds1-1/mnelson.html.

106. Z. Pawlak, Rough sets, International Journal of Computer and Information Science 11 (1982) 341-356.

107. Z. Pawlak, Rough Sets. Theoretical Aspects of Reasoning about Data, Dordrecht: Kluwer Academic, 1991.

108. Z. Pawlak, Rough sets approach to knowledge-based decision support, European Journal of Operational Research 99 (1997) 48-59.

109. Z. Pawlak, Rough sets and intelligent data analysis, Information Sciences 147 (2002) 1-12.

110. H. Pierreval, R.C. Huntsinger, An investigation on neural network capabilities as simulation metamodels, Proceedings of the 1992 Summer Computer Simulation Conference, 1992, 413-417.

111. G. Pinski, F. Narin, Citation influence for journal aggregates of scientific publications: Theory, with application to the literature of physics, Information Processing and Management 12 (1976) 297-312.

112. P. Pirolli, J., Pitkow, R. Rao, Silk from a sow's ear: Extracting usable structures from the web, Proceedings of ACM SIGCHI Conference in Human Factors in Computing, 1996.

113. J. Pitkow, P. Pirolli, Life, death, and lawfulness on the electronic frontier. Proceedings of ACM SIGCHI Conference on Human Factors in Computing, 1997.

114. J.R. Quinlan, Learning efficient classification procedures and their application to chess and games. In Machine Learning, Michalski, R.S., Carbonell, J.G., and Mitchell T.M., (Eds), Tioga Publishing Company, Palo Alto, CA, 1983.

115. D. Rocacher, On Fuzzy Bags and their application to flexible querying, Fuzzy Sets and Systems, 140(1) 93-110.

116. D. Rocacher, F. Connan, Flexible queries in object-oriented databases: on the study of bags, In Proceedings of FUZZ-IEEE '99, 8th International Conference on Fuzzy Systems, Seoul, Korea, 1999.

117. A. Schenker, H. Bunke, M. Last, A. Kandel, Graph-Theoretic Techniques for Web Content Mining, Series in Machine Perception and Artificial Intelligence, Vol. 62, World Scientific, 2005.

118. M. Schneider, E. Shnaider, A. Chew, A. Kandel, Automatic Construction of FCM, Fuzzy Sets and Systems, 9(2), 161-172, 1998.

119. H.A. Simon, Search and reasoning in problem solving, Artificial Intelligence 21 (1983) 7-29.

120. A. Skowron, C. Rauszer, The discernibility Matrices and Functions in Information Systems, Handbook of Applications and Advantages of the Rough Sets Theory, R. Slowinski (ed.), Dordrecht: Kluwer Academic, 1992, pp. 331-362.

121. R. Slowinski, J. Stefanowski, RoughDAS and Rough-class software implementation of the rough sets approach, in Intelligent Decision Support, Handbook of Applications and Advantages of the Rough Sets Theory, R. Slowinski (ed.), Dordrecht: Kluwer Academic, 1992, pp. 445-456.

122. H. Small, Co-citation in the scientific literature: A new measure of the relationship between two documents, JASIS 24 (1973) 265-269.

123. H. Small, The synthesis of specialty narratives from co-citation clusters, JASIS 37 (1986) 97-110.

124. H. Small, B.C. Griffith, The structure of the scientific literatures. I. Identifying and graphing specialties, Science Studies 4 (1974) 17-40.

125. A. Spink, J. Bateman, B.J., Jansen, Searching the web: A survey of Excite users, Journal of Internet Research: Electronic Networking Applications and Policy 9(2) (1999) 117-128.

126. L.A. Streeter, D. Vitello, A profile of drivers' map reading abilities, Human Factors 28(2) (1986) 223-239.

127. W.R. Taber, Knowledge Processing with Fuzzy Cognitive Maps, Expert Systems with Applications 2(1) (1991) 83-87.

128. E. Tufte, The Visual Display of Quantitative Information, Graphics Press, 1983.

129. C.J. Van Rijsbergen, Information retrieval, Butterworths, 1979.

130. P. Wright, Cognitive Overheads and prostheses: some issues in evaluating hypertexts, In Proceedings of the 3rd ACM Conference on Hypertext, San Antonio, 1991.

131. R.R. Yager, On the Theory of Bags, International Journal of General Systems 13 (1986) 23-37.

132. R.R. Yager, Cardinality of Fuzzy Sets Via Bags. Mathematical Modeling 9(6) (1987) 441-446.

133. S. Yu, D. Cai, J.R. Wen, W.Y. Ma, Improving pseudo-relevance feedback in web information retrieval using web page segmentation. In Proceedings of the 11th WWW Conference (WWW 12), May 20-24, 2003, Budapest, Hungary.
134. L.A. Zadeh, Fuzzy Sets. Information and Control 8(3) (1965) 338-353.
135. L.A. Zadeh, A computational approach to fuzzy quantifiers in natural languages, Comp. Math. App. 9 (1983) 149-184.

A

Graph Theory

Consider a graph G that represents a real web page and its adjacency matrix A. An entry a_{pq} in A is defined by the following:

$$a_{pq} = 1 \quad \text{if there is a link between web pages p and q.}$$
$$\quad = 0 \quad \text{Otherwise} \tag{A.1}$$

Here some of the properties that could be discovered from an adjacency matrix perspective:

1. A graph is said to be reflexive if every node in a graph is connected back to itself, i.e., $a_{pp} = 1$. The situation will happen if a page points back to itself.
2. A graph is said to be symmetric if for all edges p and q in G: iff $a_{pq} = 1$ then $a_{qp} = 1$. We say in this case that there is mutual endorsement.
3. A graph is said to be not symmetric if there exists two edges p and q in G such that iff $a_{pq} = 1$ then $a_{qp} = 0$. We say in this case that there is endorsement in one direction.
4. A graph is said to be transitive if for all edges p,q, and r: Iff $a_{pq} = 1$ and $a_{qr} = 1$ then $a_{pr} = 1$. We say in this case that all links p endorse links r even though not directly.
5. A graph is said to be anti-symmetric iff for all edges p and q: Iff $a_{pq} = 1$ then $a_{qp} = 0$
6. If two different web pages p and q point to another web page r then we say that there is social filtering. This means that these web pages are related through a meaningful relationship. $a_{pr} = 1$ and $a_{qr} = 1$.
7. If a single page p points to two different web pages q and r then we say that there is co-citation. $a_{pq} = 1$ and $a_{pr} = 1$

To illustrate the above points let us look at graph G in Figure A.1.

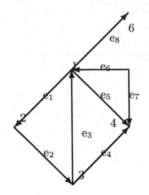

Fig. A.1. A General graph

Here are the algebraic properties of R ∈ G:

- R is not reflexive.
- R is not symmetric.
- R is not transitive.
- R is anti-symmetric. $(1,4) \in R$, $(3,4) \in R$, and $(5,4) \in R$.

we could say that the vertex with the highest number of web pages pointing to it is vertex 4. $(5,1) \in R$ and $(5,4) \in R$: 5 co-cites 1 and 4.

B

Interior Point Methods

B.1 Introduction

Definition of Interior point methods An Interior point method is a family
of algorithms that stays in the strict region within a feasible region, such as a
Barrier Function. The term grew from Karmakar's algorithm to solve a Linear
program. In that case, the resulting solution (if it exists) defines the (unique)
optimal partition.

Barrier Function P is a Barrier Function on the strict interior say I, of the
set S if P is continuous, and:

$$P(x) \to \infty$$
$$\forall x \to \text{any} S \in I.$$

This arises in penalty function methods for certain classes that include convex
programs of the form:

$$Min\{f(x) : g(x) >= 0\} \tag{B.1}$$
$$(\text{where I} = \{ \text{x: g(x)} > 0 \})$$

Two popular choices are:

1.

$$P(x) = \sum_i \frac{1}{g_i(x)} \tag{B.2}$$

2.

$$P(x) = -\sum_i \log(g_i(x)) \tag{B.3}$$

Karmarkar's algorithm An Interior point method for Linear programming which began a series of variations. The original algorithm applies to
the system:

$$Ae = 0 \quad \text{so } (e/n) \text{ is a feasible solution}$$

It further assumes the optimal objective value is known to be zero. (All of these assumptions have been relaxed in subsequent developments.)

Linear program (LP)

$$\begin{array}{lll}
\text{Optimize} & cx & \\
\text{subject to} & Ax = b \quad \text{or} &
\end{array} \qquad
\begin{array}{ll}
\text{Optimize} & cx \\
\text{subject to} & Ax \leq 0 \\
& x \geq 0
\end{array} \qquad \text{(B.4)}$$

$$\begin{array}{ll}
& x \geq 0
\end{array}$$

The standard form assumes A has full row rank. Computer systems ensure this by having a logical variable (y) augmented, so the form appears as:

$$\begin{array}{ll}
\text{Optimize} & Cx \\
\text{subject to} & Ax + y = b \\
& L \leq (x, y) \leq U
\end{array}$$

The original variables (x) are called structural. Note that each logical variable varies depending on the form of the original constraint. This computer form also represents a range constraint with simple bounds on the logical variable. Some bounds can be infinite (i.e., absent), and a free variable (logical or structural) is when both of its bounds are infinite. Optimal partition. Consider a primal-dual pair of Linear rpograms:

$$\begin{array}{lll}
\text{Minimize} & cx & \\
\text{subject to} & x \geq 0 \quad \text{or its dual form} &
\end{array} \qquad
\begin{array}{ll}
\text{Maximize} & yb \\
\text{subject to} & y \geq 0 \\
& yA \leq c
\end{array} \qquad \text{(B.5)}$$

$$\begin{array}{ll}
& Ax \geq b
\end{array}$$

We define primal surplus variables $s = Ax - b$, and dual slack variables $d = c - yA$. For any non-negative vector, v, define its support set as those indexes for which the coordinate value is positive:

$$I(v) = \{j: v_j > 0 \text{ where } v_j \text{ is associated}$$

with the jth column of the basis matrix.$\}$

In an optimal solution,

$$j \in I(x) \rightarrow j \notin I(d), \text{ and } i \in I(s) \rightarrow i \notin I(y).$$

Thus, the intersections:

$$I(x) \cap I(d) = \emptyset \ \& \ I(s) \cap I(y) = \emptyset.$$

If the solution is strictly complementary, these support sets yield partitions because then $I(x) \cup I(d) = \{1, \cdots, n\}$ and $I(s) \cup I(y) = \{1, \cdots, m\}$. A strictly complementary solution is an interior solution (and conversely), so each interior solution yields a partition of these index sets. A key fact is that every LP with an optimal solution must have an optimal interior solution. Further, every optimal interior solution yields the same partition, so it is called the optimal partition.

B.2 Applying IPM to the Backpropagation Learning Rule

Generalized backpropaation of error is essentially a gradient descent method which modifies the error by adapting weights of connections between neurons at the hidden layer and the output layer. Backpropagation provides an efficient computational method for changing the weights in a feed forward network with differentiable activation function units, to learn a training set of input/output examples. We will aply in this section IPM to backpropaation neural networks. More specifically, we expand the objective function around the current weights Linearly and minimize the approximated objective function inside a small neighborhood of the current weights through an IPM which lead us towards the optimal solution. The resulted learning rule is then applied to a given problem. We will enhance the performance of the Linear technique by considering quadric approximation of the objective function. The hessian term of the quadratic approximation of the objective function represents a heavy computational burden. We will examine the approach we are proposing in the next section. An IP algorithm for quadratic programming with box constratins will be described. And a new learning rule based on such an algorithm will be proposed.

A learning rule using inexact Hessian

The learning problem of artificial neural network can be treated within the context of nonLinear optimization. The objective function looks like the following

$$E_{avg}(w) = \frac{1}{2N} \sum_{n=1}^{N} \sum_{j \in \mathcal{O}} e_j^2(n) \tag{B.6}$$

where $\mathcal{O}$ is the output set, $e_j(n) = d_j(n) - o_j(n)$, and $d_j(n)$ and $o_j(n)$ are the desired response and actual response of the system at neuron j for pattern n at neuron j, respectively. Let $o_j(n)$ reacts according to the following nonLinear smooth function:

$$o_j(n) = \phi_j \left(\sum_{i=0}^{p} w_{ji}(n) o_i(n) \right). \tag{B.7}$$

$E_{avg}(w)$ is a function of all free parameters including the weights and the thresholds. The objective of the learning process is to adjust the free parameters of the networks to minimize $E_{avg}(w)$. The first derivative of $E_{avg}(w)$ with respect to w is:

$$\frac{\partial E_{avg}}{\partial w} = -\frac{1}{N} \sum_{n=1}^{N} \sum_{j \in O} \frac{\partial \phi_j}{\partial w} e_j(n) \tag{B.8}$$

and the second derivative or the Hessian is:

$$H = \frac{1}{N} \sum_{n=1}^{N} \sum_{j \in O} \left[\frac{\partial \phi_j}{\partial w} \cdot \frac{\partial \phi_j}{\partial w}^T - \frac{\partial^2 \phi_j}{\partial w^2} \cdot e_j(n) \right]. \tag{B.9}$$

Next we consider a network fully trained to a local minimum in error at w^*. Under this condition the output $o_i(n)$ will be close to the desired response $d_j(n)$, and hence we neglect the term involving $e_j(n)$ in the expression (B.9). This simplification, which was used by Hassibi & Stork [48], yields the following inexact Hessian:

$$Q = \frac{1}{N} \sum_{n=1}^{N} \sum_{j \in O} \frac{\partial \phi_j}{\partial w} \cdot \frac{\partial \phi_j}{\partial w}^T. \tag{B.10}$$

Then, the learning error function $E_{avg}(w)$ can be approximated by a Polynomial $P(w)$ of degree 2, in a neighborhood of w_i:

$$E_{avg}(w) \approx P(w) \equiv E_{avg}(w_i) + \nabla E_{avg}(w_i)^T (w - w_i) + \frac{1}{2}(w - w_i)^T Q(w - w_i)$$

where $\nabla E_{avg}(w_i)$ and Q are given by (B.8) and (B.10), respectively. Instead of solving a Linear programming problem for a search direction as proposed by Meghabghab and Ji in [100], we consider the following quadratic problem:

$$\begin{aligned} &\text{minimize} \quad \nabla E_{avg}(w_i)^T (w - w_i) + \frac{1}{2}(w - w_i)^T Q(w - w_i) \\ &\text{subject to} \quad -\alpha e \le w - w_i \le \alpha e, \end{aligned} \tag{B.11}$$

where e is the n–dimensional vector of all ones and $\alpha > 0$ is a constant. Let us denote $w - w_i + \alpha e$, $\nabla E_{avg}(w_i)$, and $2\alpha e$ by x, c, and b, respectively. By introducing an auxiliary vector $z = b - x$, equation (B.11) and its dual can be expressed as:

$$\begin{array}{llll} \text{minimize} \quad c^T x + \frac{1}{2} x^T Q x & & \text{maximize} \quad b^T y - \frac{1}{2} x^T Q x \\ \text{subject to} \quad x + z = b & \text{and} & \text{subject to} \quad s_x = Qx + c - y \ge 0 \\ \qquad\qquad x, z \ge & & \qquad\qquad s_z = -y \ge 0. \end{array} \tag{B.12}$$

The Kuhn-Tucker conditions imply that the duality gap $x^T s_x + z^T s_z$ is zero at optimality. That is, (x, z, y, s_x, s_z) is an optimal feasible solution if and only if:

$$\begin{cases} z = b - x, \quad x \ge 0, \quad z \ge 0, \\ s_x = Qx + c - y, \quad s_x \ge 0, \\ s_z = -y, \quad s_z \ge 0, \\ Xs_x = 0, \\ Zs_z = 0, \end{cases} \tag{B.13}$$

where $X = diag(x)$ and $Z = diag(z)$. Applying Newton's method to a perturbed system of (B.13) yields the following:

$$(x^{k+1}, z^{k+1}, y^{k+1}, s_x^{k+1}, s_z^{k+1}) = (x^k, z^k, y^k, s_x^k, s_z^k) + \theta^k (\delta x, \delta z, \delta y, \delta s_x, \delta s_z),$$
(B.14)

where the search direction is defined by:

$$\begin{cases} \delta x &= (Q + X^{-1}S_x + Z^{-1}S_z)^{-1}[Z^{-1}(Zs_z - \mu e) - X^{-1}(Xs_x - \mu e)], \\ \delta z &= -\delta x, \\ \delta s_x &= -X^{-1}(S_x \delta x + X s_x - \mu e), \\ \delta s_z &= Z^{-1}(S_z \delta x - Z s_z + \mu e), \\ \delta y &= Q\delta x - \delta s_x \ . \end{cases}$$
(B.15)

The updating scheme defined by (B.14) involves two free variables μ and θ^k. With the specific choice of μ and θ^k proposed by Han et al. [46], the interior-point path-following algorithm guarantees a polynomial complexity. In our implementation, we will use a heuristic procedure (see Procedure 1 below) suggested by Han et al. [46] to accelerate the convergence. The following point

$$x^0 = \alpha e, \ z^0 = \alpha e, \quad s_z^0 = \rho e, \ y^0 = -\rho e, \quad s_x^0 = Qx^0 + c - y^0, \quad \text{(B.16)}$$
$$\rho = \max\{\alpha, \max_{1 \le i \le n} \{\alpha - c_i - \alpha(Qe)_i\}\}, \quad \text{(B.17)}$$

can be chosen to start our procedure due to the special structure of the problem. Clearly, $(x^0, z^0, y^0, s_x^0, s_z^0)$ is a strictly feasible point to (B.12).

Procedure 1

P 1 Set $\rho_{min} = (2n)^{1.5}$ and $\beta = 0.9995$ and choose $(x^0, z^0, y^0, s_x^0, s_z^0)$ from (B.17).

P 2 For $k = 0, 1, 2, \cdots$

P 2.1 $\rho_k \longleftarrow \max\{\rho_{min}, 1/((x^k)^T s_x^k + (z^k)^T s_z^k)\}$;

P 2.2 Calculate the direction $(\delta x, \delta z, \delta y, \delta s_x, \delta s_z)$ from (B.15);

P 2.3 Set $\bar{\theta}^k = \max\{\theta | \theta \le 1, x^k + \theta\delta x \ge 0, z^k + \theta\delta z \ge 0, s_x^k + \theta\delta s_x \ge 0, s_z^k + \theta\delta s_z \ge 0\}$;

P 2.4 If $\bar{\theta}^k \le 0.5$, then divide ρ_{min} by 2 and set $\theta^k = \beta\bar{\theta}^k$;

P 2.5 If $0.5 < \bar{\theta}^k < 1.0$, then set $\theta^k = \beta\bar{\theta}^k$;

P 2.6 If $\bar{\theta}^k = 1.0$, then multiply ρ_{min} by 2 and set θ^k to 1;

P 2.7 Update the iterate $(x^k, z^k, y^k, s_x^k, s_z^k)$ from (B.14) and go to the next iteration.

A new learning rule based on *Procedure 1* can now be specified as follows.

An interior-point learning rule

A 1 Choose w_0 and $\alpha > 0$.

A 2 For $k = 0, 1, 2, \cdots$

A 2.1 Compute $\nabla E_{avg}(w_k)$, gradient of $E_{avg}(w)$ at w_k;

A 2.2 Call *Procedure* 1 and let $(x^*, z^*, y^*, s_x^*, s_z^*)$ be the solution to (B.12);

A 2.3 Update the weights by $w_{k+1} = w_k + x^* - \alpha e$.

We note that there are two loops in this algorithm. The main cost is to compute the gradient of E_{avg} at each iterate in the outer loop. Unlike the linear case where the matrix Q is zero, we need to solve a linear system of equations at each step of the inner loops. Note that there are two loops in this algorithm. The main cost is to compute the gradient of E_{avg} at each iterate in the outer loop. Unlike the linear case where the matrix Q is zero, we need to solve a linear system of equations at each step of the inner loop.

B.3 Numerical Simulation

In this section, we present some preliminary numerical experiments ([100]) on the interior-point learning rule using an inexact Hessian. The sofwtare code is written in MATLAB.

The test problem we used is the parity batch problem. The parity problem is often used for testing and evaluating network designs. It should be realized, however, that it is a very hard problem, because the output must change whenever any single input changes. This is untypical of most real-world applications problems, which usually have much more regularity and allow generalizations within classes of similar input patterns. In our example, the output is required to be 1 if the input pattern contain an odd number of 1's and 0 otherwise. We considered a parity problem with N ($N = 4$) inputs, one hidden layer of N ($N = 4$) units, which is sufficient to solve the problem of the N ($N = 4$) inputs, and one output. Only 5 patterns were used to train the neurons to allow for some generalization to happen.

812 trials were run for each one of the techniques: classic back-prop and the quadratic and Linear interior point learning rules. In all cases, the learning rate was initially set to 0.1 and then adaptively changed in relation to the sum-squared error of the objective function. The momentum was fixed at 0.95 for the standard backpropaation algorithm. To alleviate the computational complexity of each step in the inner loop, any call to *Procedure 1* and termination define an iteration if the approximated objective function has been reduced by a factor of 0.001. We also terminated a run once the number of iterations exceeds a threshold (fixed in our case to 3000) without

Table B.1. Results of the parity problem

Techniques	Percentage of Successful runs	Overall Average
QILR	100%	17.5
LILR	100%	85.6
BP	97.5%	209.6

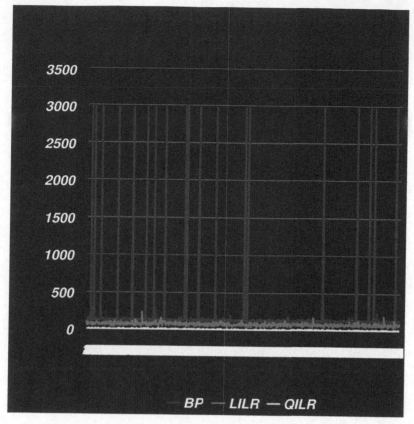

Fig. B.1. Comparing of BP, LILR, and QILR on the parity problem

any success of the sum-squared error being reduced to the error goal which is set to a value of 0.01.

From Table B.1, we could see that both quadratic interior point learning rule (QILR) and Linear interior point learning rule (LILR) were successful in solving the parity problem all the times, while classic back propagation technique (BP) was only successful 98.2% of the times. The number of iterations for all three techniques is graphed in Figure B.1; we could see that QILR always wins over LILR and classic BP. Also, LILR wins over classic BP 98.2%. Table B.1 summarizes overall average number of iterations of all three techniques. It is important ot see the reduction of the number of iterations using QILR compared to LILR and the dramatic improvement over BP (almost 15 times). This makes QILR a great new backpropagation rule that definetely ought to be implemented on other new problems (Multiple Sclerosis, and other batch problems).

C

Regression Models

Regression is a statistical tool that can be used to investigate relationships between variables. The terms "regression" and the general methods for studying relationships were introduced by in mid 1850's. regression models are used to predict one variable from one or more other variables. regression models provide researchers with a powerful tool, allowing predictions about past, present, or future events to be made with information about past or present events. Researchers employ these models either because it is less expensive in terms of time to make the predictions than to collect the information about the event itself, or, more likely, because the event to be predicted will occur in some future time. Before describing the details of the modeling process, however, some examples of the use of regression models will be presented.

C.1 Linear Regression

Linear regression attempts to model the relationship between two variables by fitting a Linear equation to observed data. One variable is considered to be an explanatory variable, and the other is considered to be a dependent variable. For example, in our case, we want to relate the time it took to search for an answer to all our twenty students successful or unsuccessful using a Linear regression model. Before attempting to fit a Linear model to observed data, one should first determine whether or not there is a relationship between the variables of interest. This does not necessarily imply that one variable causes the other but that there is some relationship between the two variables. A scatter plot can be a helpful tool in determining the strength of the relationship between two variables. If there appears to be no association between the proposed explanatory and dependent variables (i.e., the scatter plot does not indicate any increasing or decreasing trends), then fitting a Linear regression model to the data probably will not provide a useful model. A valuable numerical measure of association between two variables is the correlation coefficient, which is a value between -1 and 1 indicating the strength of the association of the observed data for the two variables. A Linear regression line has an equation of the form:

$$Y = b_0 + b_1 X \tag{C.1}$$

Where X is the explanatory variable and Y is the dependent variable. Facts about the regression line:

y-intercept: This is the value of b_0

slope: This is the value of b_1

predicted value of ith observation: $\bar{y}_i = b_0 + b_1 {}^* x_i$

error for i^{th} observation: $e_i = y_i - \bar{y}_i$

The most common method for fitting a regression line is the method of least-squares. This method calculates the best-fitting line for the observed data by minimizing the sum of the squares of the vertical deviations from each data point to the line (if a point lies on the fitted line exactly, then its vertical deviation is 0). We could choose line to make sum of errors equal to zero:

$$\sum_i e_i = e_0 + \cdots + e_n = 0 \tag{C.2}$$

But many lines satisfy this equation (C.2). Only one line satisfies:

Minimize SSE $\sum_i e_i^2 = e_0^2 + \cdots + e_n^2$

subject to $\sum_i e_i = 0$

Thus to find b_1, solve $\frac{\delta(SSE)}{\delta b_1} = 0$ for b_1.

This results in the following:

$$b_1 = \frac{\sum_{i=1}^{n}(x_i y_i - n \bar{x} \bar{y})}{\sum_{i=1}^{n}(x_i^2 - n \bar{x})}$$

$$b_0 = y - b_1 + \bar{x}$$

where $\bar{x}$ and $\bar{y}$ are the mean values of x and y.

Applying Linear Regression to the Time it took to answer a Query as a function of the number of moves In the case of Figure C.1, the 20 users of Experiment of Chapter 6 were considered. To find out whether the time it took for each user to search for the answer for the query regressed with the number of moves that it took to find the answer or exit and fail. In other terms, the more spent time to search for the answer really meant more moves that the user took.

Applying Linear regression to the Time it took to answer a Query as a function of the number of moves

From Table C.1 we can calculate average Time = 363.35, average number of Moves = 12.65, standard deviation of the number of moves = 8.511757, and

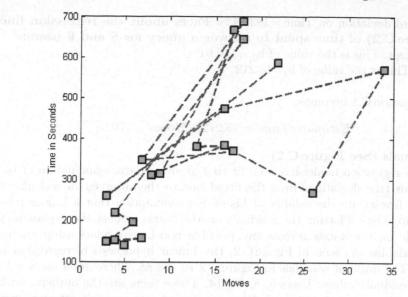

Fig. C.1. Time in Seconds as a function of the number of Moves for all 20 users

Table C.1. Summary of the 20 users

Moves	Time
22	587
3	258
16	473
34	569
26	267
17	369
7	346
18	689
9	314
8	310
16	384
13	380
17	667
18	645
4	219
6	195
3	148
7	157
4	151
5	139

standard deviation on Time $= 183.8438$. **Facts about the regression line (Figure C.2) of time spent to answer a query for S and F users:**
y-intercept: This is the value of $b_0 = 170.91$
slope: This is the value of $b_1 = 15.212$

Equation C.1 becomes:

$$EstimatedTime = 15.212 * Moves + 170.91 \qquad (C.3)$$

Residuals (See Figure C.2)

Once a regression model has been fit to a group of data, examination of the residuals (the deviations from the fitted line to the observed values) allows one to investigate the validity of his or her assumption that a Linear relationship exists. Plotting the residuals on the y-axis against the explanatory variable on the x-axis reveals any possible non-Linear relationship among the variables. As seen in Figure C.2, the Linear hypothesis of regression is weak. The norm of residuals in Figure C.2 is: 568.88. There are 4 users with large residuals values: Users 5, 8, 13, 14. These users are the outliers. Such points may represent erroneous data, or may indicate a poorly fitting regression line.

Correlation Coefficient

The strength of the Linear association between two variables is quantified by the correlation coefficient. Given a set of observations (x_1, y_1), $(x_2, y_2), \cdots,$ (x_n, y_n), the formula for computing the correlation coefficient is given by:

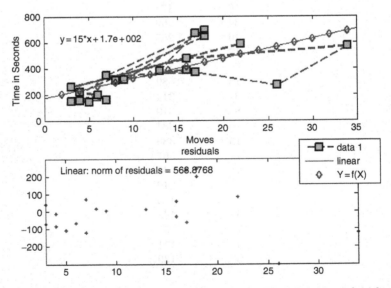

Fig. C.2. Linear Regression of Time with residuals. Users 5,8,13, and 14 have the largest residuals values

$$r = \frac{1}{n} \sum_{i}^{n} \frac{(x - \bar{x})(y - \bar{y})}{s_x s_y} \tag{C.4}$$

Where $\bar{x}$ and $\bar{y}$ are the mean values of x and y, and s_x and s_y are the standard deviation of x and y.

The correlation coefficient always takes a value between -1 and 1, with 1 or -1 indicating perfect correlation (all points would lie along a straight line in this case). A positive correlation indicates a positive association between the variables (increasing values in one variable correspond to increasing values in the other variable), while a negative correlation indicates a negative association between the variables (increasing values is one variable correspond to decreasing values in the other variable). A correlation value close to 0 indicates no association between the variables. Since the formula for calculating the correlation coefficient standardizes the variables, changes in scale or units of measurement will not affect its value. For this reason, the correlation coefficient is often more useful than a graphical depiction in determining the strength of the association between two variables. It turns out that r in this case using Table C.1 is r = 0.70431.

Correlation in Linear regression

The square of the correlation coefficient, r^2, is a useful value in Linear regression. This value represents the fraction of the variation in one variable that may be explained by the other variable. Thus, if a correlation of 0.70431 is observed between two variables (say, time and moves, for example), then a Linear regression model attempting to explain either variable in terms of the other variable will account for 49.6% of the variability in the data. The correlation coefficient also relates directly to the regression line Y = a + bX for any two variables, where:

$$b = \frac{r s_y}{s_x} \tag{C.5}$$

Std on Moves = s_x = 8.511757, and Std on Time = s_y = 183.8438, thus: b=15.2128 (slope of C.3)
Because the least-squares regression line will always pass through the means of x and y, the regression line may be entirely described by the means, standard deviations, and correlation of the two variables under investigation.

Matrix Interpretation of Linear regression

A matrix interpretation X_1 of the above equation also called a design matrix is:
And using P for X_1 the regression equation would be:

$$Y = bP \tag{C.6}$$

$$X_1 = \begin{bmatrix} 1 & 22 \\ 1 & 3 \\ 1 & 16 \\ 1 & 34 \\ 1 & 26 \\ 1 & 17 \\ 1 & 7 \\ 1 & 18 \\ 1 & 9 \\ 1 & 8 \\ 1 & 16 \\ 1 & 13 \\ 1 & 17 \\ 1 & 18 \\ 1 & 4 \\ 1 & 6 \\ 1 & 3 \\ 1 & 7 \\ 1 & 4 \\ 1 & 5 \end{bmatrix}$$

where $b = [b_0 \ b_1]^T$. Solving for b yields:

$$b = YP^{-1} \tag{C.7}$$

using for P the values of Moves and Y the values of Time, we get b=[170.9148, 15.2123], which correspond respectively to the Y intercept and the slope of equation C.3.

C.2 Polynomial Regression

If the simple regression design is for a higher-order effect of P, say the seventh degree effect, the values in the X_1 column of the design matrix would be raised to the 7th power and using all the terms from P_1 to P_7 for X_1 the regression equation would be but do not include interaction effects between predictor variables:

$$Y = b_0 + b_1 P + b_2 P^2 + b_3 P^3 + b_4 P^4 + b_5 P^5 + b_6 P^6 + b_7 P^7 \tag{C.8}$$

In regression designs, values on the continuous predictor variables are raised to the desired power and used as the values for the X variables. No recoding is performed. It is therefore sufficient, in describing regression designs, to simply describe the regression equation without explicitly describing the design matrix X. In matrix terms this leads to:

$$Y = bP' \tag{C.9}$$

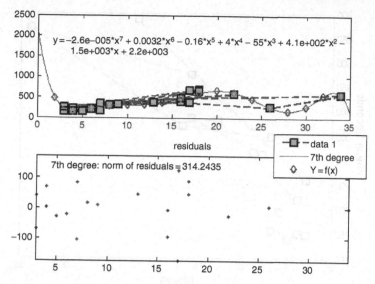

Fig. C.3. Seventh Polynomial regression of Time with residuals as a function of Moves.

Where $b = [b_0\ b_1\ b_2\ b_3\ b_4\ b_5\ b_6\ b_7]$ and $P' = [1\ P\ P^2\ P^3\ P^4\ P^5\ P^6\ P^7]$
Solving for b yields:

$$b = YP'^{-1} \qquad\qquad (C.10)$$

using for P the values of Moves and Y the values of Time, we get b=[2174, −478.4,408.4,−55,4,−.2,0,−0], which correspond respectively to the values below in the middle of the next page.

Figure C.3 shows a seventh degree regression Polynomial of the Time with all the parameters $[b_0\ b_1\ b_2\ b_3\ b_4\ b_5\ b_6\ b_7]$ as follows: b_0=2174, b_1=−1478.4, b_2=408.44, b_3=−55.015, b_4=4, b_5=−0.15869, b_6=0.00321, and b_7=−0.000026. The norm of residuals is 314.3425. Note the 45% drop of the norm of the residuals as compared with the Linear norm from 568.88 to 314.3425.

Linear regression after Removing out the Outliers:
If users 5, 8, 13, and 14 were removed, (see figure C.4 for the actual time without outliers), figure C.5 shows the new Linear fitting with a larger slope and a smaller Y intercept. Slope=15.126 Y intercept: 147.95
Equation C.1 becomes:

$$EstimatedTime = 15.126Moves + 147.95 \qquad (C.11)$$

With a norm of residuals = 261.99.
Comparing the latter with the norm of residuals before removing the outliers: 568.88 is a 54% drop.

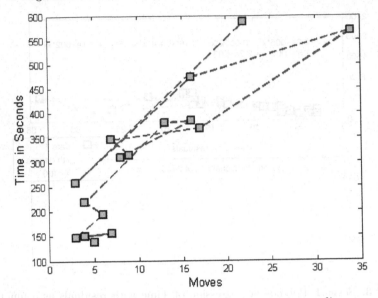

Fig. C.4. Actual Time in seconds after removing outliers

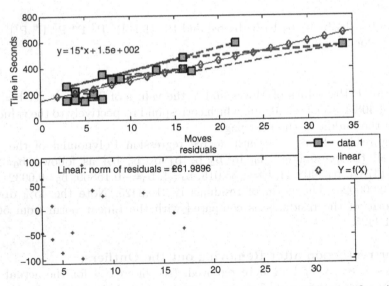

Fig. C.5. linear regression analysis of Time after removing outliers with residuals.

By applying equation C.9 also we get the following results: b = [147.9465 15.1256] which are exactly the Y intercept and the slope of equation C.10. If a point lies far from the other data in the horizontal direction, it is known as an **influential observation**. The reason for this distinction is that these

Table C.2. Summary of time and Moves after removing users: 5, 8, 13, and 14

Moves	Time
22	587
3	258
16	473
34	569
17	369
7	346
9	314
8	310
16	384
13	380
4	219
6	195
3	148
7	157
4	151
5	139

points have may have a significant impact on the slope of the regression line. Notice that no user 2 fits that description.

Correlation coefficient after removing Outliers:
Since the outliers are removed, the correlation coefficient must increase. An application of C.4 applied to Table C.2 yields:

$$r = 0.883908$$

r is getting closer to 1 indicating almost a perfect correlation and a positive association. Thus, if a correlation of 0.883908 is observed between two variables (say, time and moves, for example), then a Linear regression model attempting to explain either variable in terms of the other variable will account for 78.13% of the variability in the data. That is a 3/2 times more accounting of the variability of the data correlation than what is used to be before removing the outliers. The correlation coefficient also in C.5 can be applied after removing outliers

$$b = \frac{r s_y}{s_x} \tag{C.12}$$

Std on Moves = s_x = 8.45, and Std on Time = s_y = 144.65, thus: b = 15.1256 (slope of (9))

Polynomial regression after Removing the Outliers:
The equation of Polynomial regression C.8 is applied to the data in Table C.2.

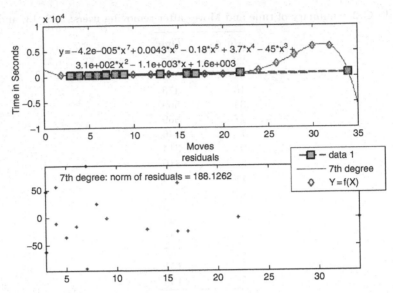

Fig. C.6. A seventh polynomial regression of Time after removing outliers with residuals

Figure C.6 shows a seventh degree regression Polynomial of the Time as a function of the number of moves with all the parameters [b_0 b_1 b_2 b_3 b_4 b_5 b_6 b_7] as follows: $b_0 = 1613.3$, $b_1 = -1081.8$, $b_2 = 311.64$, $b_3 = -45.434$, $b_4 = 3.75$, $b_5 = -0.17$, $b_6 = 0.0043$, and $b_7 = -4.2 * 10 - 5$. The norm of residuals is 188.1262. Note the 36% drop of the norm of the residuals as compared with the Linear norm from 261.9896 to 188.1262. We get the same results also by applying equation C.3 and using for P the values of Moves and Y the values of Time, we get $b = [1613.3, -1081.8, 311.6, -45.4, 3.7, -.2, 0, -0]$, which correspond respectively to the values above in the previous page.

C.3 Multiple Regression

The regression that we performed so far involved 2 variables: the number of moves and the time it took to get an answer for the query. The number of moves does not seem seem solely to explain the time it took to answer a query. What if we added the number of web searches that students did as another independent variable. This the variable Time is a function of more than one independent variable, the matrix equations that express the relationships among the variables can be expanded to accommodate the additional data. (See figure C.7). The general purpose of Multiple regression (the term was first used by Pearson, 1908) is to analyze the relationship between several independent or predictor variables like Moves and the number of web searches, and a

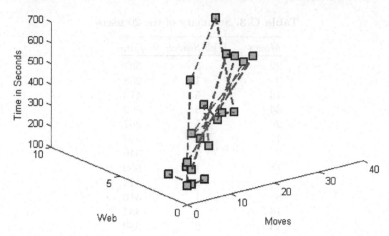

Fig. C.7. A 3-D plot of Actual Time in seconds as a function of number of Moves and the number of Web Searches

dependent or criterion variable like Time. The regression equation for a Multiple regression design for the first-order effects of two continuous predictor variables P, and Q would be:

$$Y = b_0 + b_1 P + b_2 Q. \tag{C.13}$$

Where P is Moves like in equation C.6 and Q is the number of web searches. A multivariate model of the data is an application of equation C.13:

$$Time = b_0 + b_1 Moves + b_2 Web \tag{C.14}$$

Multiple regression solves for unknown coefficients, b_0, b_1, and b_2, by performing a least squares fit. Rewriting equation C.14 as:

$$Y = bP \tag{C.15}$$

And solving for b in equation C.15 where b= $[b_0, b_1, b_2]$ yields:

$$b = YP^{-1} \tag{C.16}$$

The data in table 3 yields a vector b=[183.7006, 26.2537,-58.6386]; Table C.3 summarizes the data already collected in the experiment of 20 users (Chapter 6). To validate the model, we find the absolute value of the deviation of the data from the model (see Equation C.8):
Estimatedtime = $P * b$;
Norm of Residuals= $\|(EstimatedTime - Time)\|$= 483.9862

C.4 Factorial Regression Analysis of Time

Factorial regression designs are similar to Factorial ANOVA designs, in which combinations of the levels of the factors are represented in the design. In Factorial regression designs, however, there may be many more such

Table C.3. Summary of the 20 users

Moves	Web Search	Time
22	3	587
3	1	258
16	5	473
34	9	569
26	5	267
17	3	369
7	2	346
18	1	689
9	2	314
8	1	310
16	4	384
13	2	380
17	2	667
18	3	645
4	1	219
6	2	195
3	1	148
7	1	157
4	1	151
5	3	139

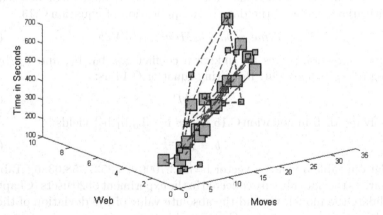

Fig. C.8. A multiple regression analysis of Time: Large square=regressed time, Small square=actual time.

possible combinations of distinct levels for the continuous predictor variables than there are cases in the data set. To simplify matters, full-Factorial regression designs are defined as designs in which all possible products of the continuous predictor variables are represented in the design. For example, the full-Factorial regression design for two continuous predictor variables P and

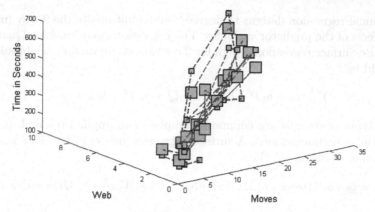

Fig. C.9. A factorial time regression analysis of time: Large square=regressed time, smaller square=actual time.

Q would include the main effects (i.e., the first-order effects) of P and Q and their 2-way P by Q interaction effect, which is represented by the product of P and Q scores for each case. The regression equation would be:

$$Y = b_0 + b_1 P + b_2 Q + b_3 P * Q \tag{C.17}$$

A Factorial model of the data is an application of equation C.17:

$$Time = b_0 + b_1 Moves + b_2 Web + b_3 * Moves * Web \tag{C.18}$$

Multiple regression solves for unknown coefficients, b_0, b_1, b_2, and b_3 by performing a least squares fit. Rewriting equation C.18 as:

$$Y = bP \tag{C.19}$$

And solving for b in equation C.19 where $b = [b_0, b_1, b_2, b_3]$ yields:

$$b = YP^{-1} \tag{C.20}$$

b = [123.264928.7928 − 25.5442 − 1.2591] To validate the model, we find the absolute value of the deviation of the data from the model. (See figure C.9)
Estimatedtime = P*b;
Norm of residuals= $\|(EstimatedTime - Time)\|$= 471.8558

C.5 Quadratic Response Surface Regression

Quadratic Response Surface regression designs are a hybrid type of design with characteristics of both Polynomial designs and fractional regression designs. Quadratic Response Surface regression designs contain all the same effects of

Polynomial regression designs to degree 2 and additionally the 2-way interaction effects of the predictor variables. The regression equation for a Quadratic Response Surface regression design for 2 continuous predictor variables P, and Q would be :

$$Y = b_0 + b_1 P + b_2 P^2 + b_3 Q + b_4 Q^2 + b_5 P * Q \tag{C.21}$$

These types of designs are commonly employed in applied research (e.g., in industrial experimentation). A surface regression model of the data is an application of equation (C.21):

$$Time = b_0 + b_1 Moves + b_2 Moves^2 + b_3 Web + b_4 Web^2 + b_5 Moves Web \tag{C.22}$$

Surface regression solves for unknown coefficients, b_0, b_1, b_2, b_3, b_4, and b_5, by performing a least squares fit. Rewriting equation C.21 as:

$$Y = bP \tag{C.23}$$

And solving for b in the above equation C.23 where b= $[b_0, b_1, b_2, b_3, b_4, b_5]$ yields:

$$b = YP^{-1} \tag{C.24}$$

b $= [79.631, 56.1669, -.3352, -108.5102, 33.8725, -9.239]$. To validate the model, we find the absolute value of the deviation of the data from the model (see figure C.10) EstimatedTime = P*b;
Norm of Errors between both time = $||(EstimatedTime - Time)|| = 381.1687$

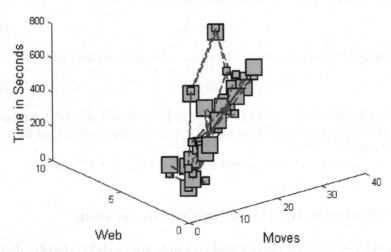

Fig. C.10. A Response Surface time regression model of time: Large square=Surface regressed time, smaller square=actual time.of time

Multiple regression after Removing Outliers

The Multiple regression that we performed in equations C.13 through C.16 revealed 4 users with the highest residuals: they are users 5, 13, 14, and 18. We already experienced better results when we removed outliers with Linear and Polynomial regression analysis above. (see figure C.11). Table C.4 summarizes the data already collected in the experiment of 16 users left after removing the outliers (Chapter 6). Application of equations C.13 through C.16 reveal

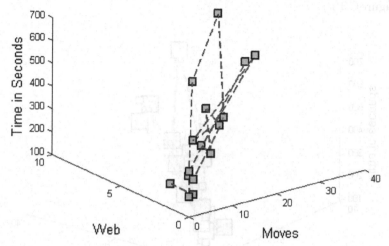

Fig. C.11. a 3-D plot of Actual Time in seconds as a function of number of Moves and the number of Web Searches after removing outliers.

Table C.4. Summary of the 16 users after removing the outliers.

Moves	Web Search	Time
22	3	587
3	1	258
16	5	473
34	9	569
17	3	369
7	2	346
18	1	689
9	2	314
8	1	310
16	4	384
13	2	380
4	1	219
6	2	195
3	1	148
4	1	151
5	3	139

a vector b = [171.2488,26.5098,-51.5438]. To validate the model, we find the norm value of the deviation of the data from the model. (see figure C.12)
Estimatedtime = P*b;
Norm of Residuals = $||norm(EstimatedTime - Time)|| = 247.367$

Factorial regression Analysis of Time after removing Outliers
Application of equations C.17 through C.20 to the users in table E.1 reveal a vector b = [124.8455,28.6967,-27.6502,-.9274]
(see figure C.13).

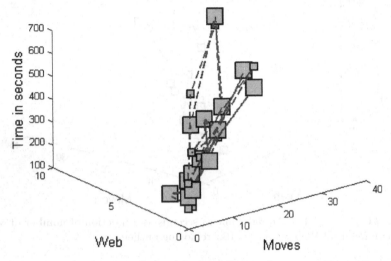

Fig. C.12. A multiple regression analysis of Time after removing outliers: Large square=regressed time, small square=actual time.

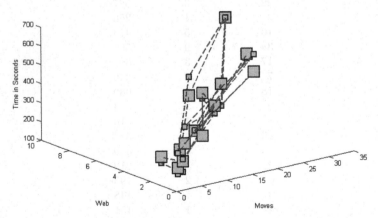

Fig. C.13. A factorial time regression analysis of time after removing the outliers. Large square=factorial regressed time, smaller square=actual time.

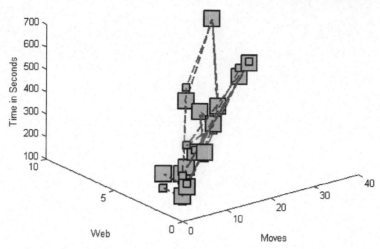

Fig. C.14. A Response Surface time regression model of time after removing the outliers. Large square=Surface regressed time, smaller square=actual time.

Response Surface regression after Removing outliers
Application of equations C.21 through C.24 to the users in table C.4 reveals a vector b=[123.2847,39.7677,.0774,-74.1755,23.3407, -7.2246] To validate the model, we find the norm value of the deviation of the data from the model. (see figure C.14)
Estimatedtime = P*b;
Norm of Residuals= $||(EstimatedTime - Time)||$= 194.2069

C.6 Final Observations on Regression Models

The Quadratic Response Surface time regression model proves to be the best model. It takes into consideration to a degree 2 the effects of Polynomial regression designs and additionally the 2-way interaction effects of the predictor variables. It reduced the norm of the error between the regressed time value and the actual time by at least 22% compared to Multiple regression before the outliers were removed. It reduces also the norm of errors by 22% after the outliers were removed compared to Multiple regression. It is the best statistical regression model if more than 1 predictor variable is to be used. In Chapter 5 we used the same technique and compared its results to Radial Basis functions with Interior point methods and without Interior point methods.

D

Fuzzy Set Theory and Fuzzy Similarity Measures

D.1 The Emergence of Fuzzy Set Theory

We live in a world where we are always making decisions, and yet these decisions are information based. Often times, we do not have a complete information at every crossroad. The lack of information or partial information that we have available to us, constitute what we call uncertainty. The seminal work of L. Zadeh [[134]] marks the beginning of the emergence of the concept of uncertainty distinct from the concept of probability. Zadeh [[134]] defines a fuzzy set as a set that has no sharp or precise boundaries. If we look at the set of synonyms to the term "probable" as given by the thesaurus provided by the software Microsoft Word we find that the probable=likely,credible,possible,feasible,plausible, apparent. If we were to rely on the "Word" 's thesaurus only, then these terms constitue a precise set with sharp boundaries. If we were to look up the synonyms of plausible for example, we find for example that "believable" is a synonym to plausible. Although "believable" is synonym of plausible and plausible is a synonym of Probable (see Figure D.1), according to the sharp set of synonyms of Probable, "believable" is not a member of the set. Instead of looking at these terms as definetly belonging or not belonging to a set, a member of a fuzzy set belongs to a fuzzy set with a certain degree of membership. Let us reframe the above example in terms of similarity ([95]): "Probable" is similar to "plausible" and "plausible" is similar to "believable". According to the Word's thesaurus: "probable" is not similar to "believable". In what follows we will use the idea of a bow tie graphs to solve such an issue. A bow tie graph is a graph such that at the center of the bow tie is the knot, which is "the strongly connected core", the left bow consists of pages that eventually allow users to reach the core, but that cannot themselves be reached from the core, i.e., L, where L=$(L_1,L_2,\cdots,L_m)$, and the right bow consists of "termination" pages that can be accessed via links from the core but that do not link back into the core, i.e., R where R = $(R_1,R_2,\cdots,R_n)$, and where n is different from m. We use this idea to construct a new Fuzzy Similarity Measure between 2 fuzzy concepts. We say that

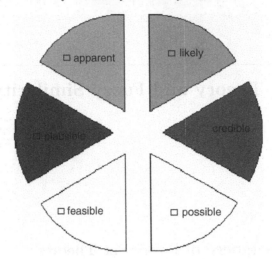

Fig. D.1. A precise sharp set for the synonyms of "Probable"

2 fuzzy terms t_1 and t_2 are fuzzy ϵ ($\epsilon \geq 0$) "Bow Tie similar" or "$\approx_\epsilon$ BT" iff:

$$t_1 \approx_\epsilon BT t_2 \text{ iff} |(C_{t_1,t_2})| \geq \epsilon \text{ and} |C_{t_2,t_1}| \geq \epsilon \qquad \text{(D.1)}$$

where C_{t_i,t_j} is the centrality score distribution of term t_i for term t_j. A centrality score C_{t_i} is defined iteratively as:

$$(C_{t_i})_k = 1 \text{ at } k = 0 \qquad \text{(D.2)}$$

$$(C_{t_i})_{k+1} = Max\left(Min\left(\frac{B \circ (C_{t_i})_k \circ A^T, B^T \circ (C_{t_i})_k \circ A))}{|B \circ (C_{t_i})_k \circ A^T + B^T \circ (C_{t_i})_k \circ A)|} \right) \right) \qquad \text{(D.3)}$$

where: $\circ$ is the symbol of composition of 2 relations, T is the transpose of the mapping, | is an absolute value, B is a fuzzy relation on L × {t1, t2}, *where* $L = (L_1, L_2, \cdots,$ and $L_m)$ (See above). A is a fuzzy relation on {t1, t2}× R, *where* $R = (R_1, R_2, \cdots, R_n)$ (See above).

Remember that the centrality score definition stems from the following observation on bow tie graphs:

$$(L_1, L_2, \cdots, L_m) \text{ is } t_1$$
$$(L_1, L_2, \cdots, L_m) \text{ is } t_2$$
$$t_1 \text{ is } (R_1, R_2, \cdots, R_n)$$
$$t_2 \text{ is } (R_1, R_2, \cdots, R_n)$$

Our Fuzzy Similarity Measure is a new Fuzzy Similarity Measure. We apply (D.1) to terms that we picked this time from an online dictionary at http://www.eleves.ens.fr/home/senellar/. The results show that similarity between terms based on fuzzy ϵ "Bow Tie similar" is a better similarity between terms than the similarity provided by Microsoft Word's Thesuarus.

D.2 Extending Blondels' Measure

In chapter 3, section 3.4, we considered Blondel's Similarity Procedure. Blondel's and VanDooren [9] applied such a similarity to the extraction of synonyms between terms. Fig. D.2 illustrates the graph of "likely" as compared to "probable". Fig. D.3 illustrates a 3-D map of the centrality scores of the nodes {1,2,3,4,5,6,7,8,9} as illustrated in the procedure Similarity as they correspond to {invidious, truthy, verisimilar, likely, probable, adapted, giving, belief, probably}.

The peak correspond to the fact that likely is first very similar to itself followed by probable, and that probable is first very similar to itself and then to likely. One natural extension of the concept of centrality is to consider networks rather than graphs; this amounts to consider adjacency matrices with arbitrary real entries and not just integers. The definitions and results presented in ([9]) use only the property that the adjacency matrices involved have non-negative entries, and so all results remain valid for networks with non-negative weights. The extension to networks makes a sensitivity analysis possible: How sensitive is the similarity matrix to the weights in the network? Experiments and qualitative arguments show that, for most networks, similarity scores are almost everywhere continuous functions of the network entries. These questions can probably also be related to the large literature on eigenvalues and invariant subspaces of graphs ([23]). We consider fuzzifying the adjacency matrices. We asked experts in linguistics to provide us with a scale of 0 to 1 instead of 1 for all terms that make Fig. D.2. Fig. D.4 is a corresponding fuzzy adjacency matrix of Fig. D.2. We apply the same Similarity procedure to the relations of A and B. Here is the final result of the centrality score CS:

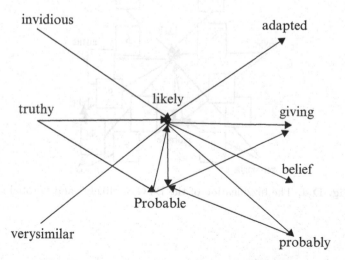

Fig. D.2. The BTG of "likely" and "probable"

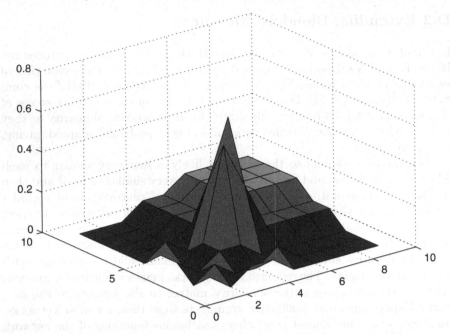

Fig. D.3. A mesh of the centrality Matrix of Likely and probable

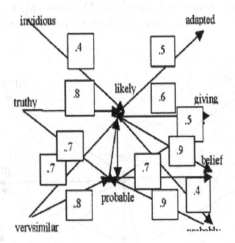

Fig. D.4. The fuzzification of the BTG of "likely" and "probable"

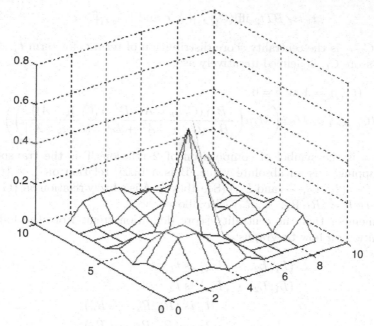

Fig. D.5. A mesh of the centrality matrix corresponding to Fig. D.4

	I	T	V	L	P_e	A	G	B	P_y
I	.02	.05	.04	0	.04	0	0	0	0
T	.05	.17	.17	.08	.07	0	0	0	.1
V	.04	.17	.17	.1	.06	0	0	0	.11
L	0	.08	.1	.59	.31	0	.05	.06	.11
P_e	.04	.07	.06	.31	.56	.06	.07	.11	.05
A	0	0	0	0	.06	.04	.05	.07	.03
G	0	0	0	.05	.07	.05	.09	.14	.04
B	0	0	0	.06	.11	.07	.14	.19	.05
P_y	0	.1	.11	.11	.05	.03	.04	.05	.15

The mesh of Fig. D.5 shows more peaks than the one in Fig. D.3, but no major changes did occur. It shows that likely and probable are very similar because of the peak between nodes 4 and 5. The fuzzification of the adjacency matrix did not improve or change the interpretation of the results of the Similarity procedure above.

D.3 The New Fuzzy Centrality Score

The procedure Similarity above has not been fuzzified yet. We use this idea to construct a new Fuzzy Similarity Measure between 2 fuzzy concepts. We say that 2 fuzzy terms t_1 and t_2 are fuzzy ϵ ($\epsilon \geq 0$) "Bow Tie similar" or "$\approx_\epsilon$ BT" iff:

$$t_1 \approx_\epsilon BTt_2 \text{ iff} |(C_{t_1,t_2})| \geq \epsilon \text{ and } |C_{t_2,t_1}| \geq \epsilon \qquad (D.4)$$

where C_{t_i,t_j} is the centrality score distribution of term t_i for term t_j. A centrality score C_{t_i} is defined iteratively as:

$$(C_{t_i})_k = 1 \text{ at } k = 0 \qquad (D.5)$$

$$(C_{t_i})_{k+1} = Max\left(Min\left(\frac{B \circ (C_{t_i})_k \circ A^T, B^T \circ (C_{t_i})_k \circ A))}{|B \circ (C_{t_i})_k \circ A^T + B^T \circ (C_{t_i})_k \circ A)|}\right)\right) \qquad (D.6)$$

where: $\circ$ is the symbol of composition of 2 relations, T is the transpose of the mapping, $|$ is an absolute value, B is a fuzzy relation on L $\times$ $\{t1, t2\}$, where $L = (L_1, L_2, \cdots, \text{and } L_m)$ (See above). A is a fuzzy relation on $\{t1, t2\} \times$ R, where $R = (R_1, R_2, \cdots, R_n)$ (See above).

Remember that the centrality score definition stems from the following observation on bow tie graphs:

$$(L_1, L_2, \cdots, L_m) \text{ is } t_1$$
$$(L_1, L_2, \cdots, L_m) \text{ is } t_2$$
$$t_1 \text{ is } (R_1, R_2, \cdots, R_n)$$
$$t_2 \text{ is } (R_1, R_2, \cdots, R_n)$$

We apply the same Procedure Similarity to Fig. D.4. Here is the final result of the fuzzy centrality score after a few iterations on "likely" and "probable":

	I	T	V	L	P_e	A	G	B	P_y
I	.095	.095	.095	0	.095	0	0	0	0
T	.095	.19	.166	.166	.143	0	0	0	.167
V	.095	.166	.19	.19	.143	0	0	0	.19
L	0	.166	.19	.21	.167	0	.119	.14	.19
P_e	.095	.143	.143	.166	.214	.119	.143	.19	.095
A	0	0	0	0	.119	.119	.119	.119	.095
G	0	0	0	.119	.143	.119	.143	.143	.095
B	0	0	0	.143	.19	.119	.143	.214	.095
P_y	0	.166	.19	.19	.095	.095	.095	.095	.214

New similarities were discovered among other terms. The algorithm had a purpose of comparing the similarity of "likely" to "probable" or the terms of the middle. Instead, other similarities were uncovered as well by looking at different α cuts:

For α=0.1664:
truthy = {.19/truthy,.1664/verisimilar,.1664/likely,.1664/probably}
verisimilar = {.1664/truthy,.19/verisimilar,.19/likely,.19/probably}
likely = {.1664/truthy,.1902/verisimilar,.214/likely,.1664/probable,.19/
        probably}
probable = {.1664/likely,.214/probable,.19/belief}

belief = {.19/probable,.214/belief}
probably = {.19/verisimilar,.19/likely,.214/probably}
likely ~ verisimilar ~ probably
probably ~ verisimilar ~ likely
verisimilar ~ likely ~ probably
truthy ~ verisimilar ~ likely ~ probably
probable ~ belief
adapted ~ giving ~ belief
giving ~ adapted ~ belief
belief ~ probable
where ~ standsforsimilar.
Forα = 0.19
truthy = {.1902/truthy}
verisimilar = {.1902/verisimilar,.1902/likely,.1902/probable}
likely = {.1902/verisimilar,.214/likely,.1902/probably}
probable = {.214/probable,.1902/belief}
belief = {.1902/probable,.214/belief}
probably = {.1902/verisimilar,.1902/likely,.214/probably}
verisimilar ~ likely ~ probably
likely ~ verisimilar ~ probably
probably ~ verisimilar ~ likely
probable ~ belief
belief ~ probable
where ~ standsforsimilar.

Fuzzy centrality score(FCS) is not reflexive. FCS is symmetric and not transitive. Figure D.6 shows the mesh of FCS. To make FCS transitive we

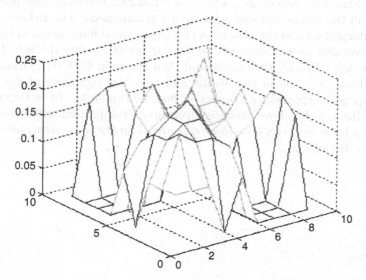

Fig. D.6. A mesh of the fuzzy centrality matrix corresponding to Fig. D.4

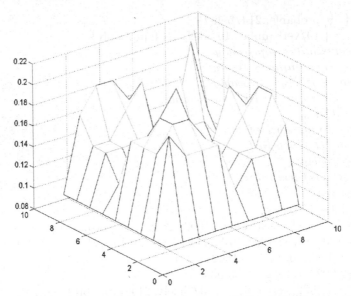

Fig. D.7. A mesh of the transitive FCS of Fig. D.6

added the following step:

$$FCS_{transitive} = FCS \bigcup (FCS \circ FCS) \qquad (D.7)$$

Figure D.7 shows the mesh of the FCS made transitive.
We started by looking at the similarity between graphs and ultimately its application to the extraction of two synonymous terms. We extended the simple "Similarity procedure", which once fuzzified revealed more properties among all the terms that make the terms synonymous. The richness of the fuzzy interpretation of the algorithm of [9], uncovered more peaks in the mesh of its centrality score. Some of the similar terms involve the set of terms from the left CBG(L), the middle and then the right CBG(R). Other terms involved only the one in the middle and then the right side of the graph. This appendix explored the concept of bow tie similarity. To further exploit the goodness of the above similarity measure, a comparison with other fuzzy similarity between terms need to be considered on the same terms taken from the same dictionary.

E

Information Theory

Information Theory provides a mathematical basis for measuring the information content of a message. We may think of a message as an instance of a message in a universe of possible messages. Shannon formalized this by defining the amount of information in a message as a function of the probability of occurrence p of each possible message, namely $-\log_2 p$. Given a universe of messages, $M = \{m_1, m_2, \cdots, m_n\}$ and a probability $p(m_i)$, for the document of each message, the expected information content of a message M is given by:

$$M = -\sum_i p(m_i) \log_2 p(m_i) \qquad \text{(E.1)}$$

The information in a message is measured by bits. The information content of a message telling the outcome whether a web page is the answer of a query q or not is:

$|\text{Web page}| = -p(\in q) \log_2(p \in q) - p(\notin q) \log_2(p \notin q)$
$|\text{Web page}| = -(1/2) \log_2(1/2) - (1/2)(\log_2(1/2) = -2(1/2) * \log_2(1/2)$
$|\text{Web page}| = -1 * \log_2(1/2)$
$|\text{Web page}| = 1 \text{bit}$

How does "popularity" change the outcome of whether a web page will be considered or not as part of the answer of a query or not? Let us assume that a web page can be rigged so that it has a chance of coming out as part of an answer to a query 25%. The information content of a message telling the outcome whether a web page is the answer of a query or not is:

$|\text{Web page}| = -(1/4) * log_2(1/4) - (3/4) log_2(3/4)$
$|\text{Web page}| = .5 + (3/4)*.415$
$|\text{Web page}| = .81128$

E.1 Application of Information Theory to a Production Line Problem

The following example is based on data taken from a paper by Gento and Redondo ([40]). The maintenance of production systems depends on large number of variables and usually incorporates high degree of uncertainty. For these aspects, the decisions are very complex. The size of maintenance database has increased lot in the last few years due to the use of programmable logic controllers and data acquisition boards. The maintenance manager needs to forecast the system state based on the system load, maintenance and production scheduling and so on. The application chosen to illustrate the power of Information Theory involves an area recently automated in the assembly line of a product that depends on various elements (steady and movable) to which PVC compound has to be applied. The complete line has six main tables and two auxiliary ones only used as links, as well as three robots with three degrees of freedom plus an additional degree of freedom given in rail over which they can move. Four attributes are considered (one for each type of machine): A_1-rail, A_2-robot, A_3-main table and A_4-auxiliary table. A_1-rail can have two different states S_1 and S_2; and A_2-robot, A_3-main table and A_4-auxiliary table can have three different states S_1, S_2 and S_3. To maintain data confidentiality we do not specify these states. The possible outputs of the complete system are shown in Table E.1. Also, the data supplied by the company expert is an incomplete set of data2 (see Table E.2), but it is not problem because lot of situations can only be analyzed by considering an incomplete set of cases. This situation is very possible in the maintenance of production systems where we cannot get all different combinations of variables.

When all attributes were checked whether relevant to the decision classification, removing A_4 meant the decision classification was still intact. Table E.3 was deduced by that taking into consideration the latter fact and by eliminating all redundant examples from the table. We first calculate the entropy of each attribute knowing that since we have 4 possible outputs, the logarithm used is a log based 4 and not logarithm based 2.

A_1:H $= (6/9)(-(2/6) * \log_4(2/6) - (2/6) * \log_4(2/6) - (2/6) * \log_4(2/6)$
A_1:H $= (2/3)((-6/6) * \log_4(2/6))$
A_1:H $= -(2/3) * \log_4(1/3) = 0.528321$

Table E.1. Summary of the outputs

Outputs	Corresponding Meaning
W	*Working*
A	*Alarm*
F_1	*Fault of type 1 (light)*
F_2	*Fault of type 2 (critical)*

Table E.2. Data Supplied by the expert's company

Inputs	Attributes				Outputs
	A_1	A_2	A_3	A_4	
x_1	S_1	S_3	S_2	S_2	W
x_2	S_1	S_3	S_3	S_2	W
x_3	S_1	S_3	S_3	S_3	W
x_4	S_1	S_2	S_2	S_2	A
x_5	S_1	S_2	S_2	S_3	A
x_6	S_1	S_2	S_3	S_2	A
x_7	S_1	S_1	S_1	S_3	F_1
x_8	S_1	S_3	S_1	S_2	F_1
x_9	S_1	S_3	S_1	S_3	F_1
x_{10}	S_2	S_1	S_1	S_2	F_2
x_{11}	S_2	S_1	S_1	S_3	F_2
x_{12}	S_2	S_1	S_3	S_1	F_2
x_{13}	S_2	S_1	S_3	S_2	F_2
x_{14}	S_2	S_1	S_3	S_3	F_2
x_{15}	S_2	S_2	S_1	S_1	F_2

Table E.3. Set of examples with indispensable attributes

Inputs	Inputs'	Attributes			Outputs
		A_1	A_2	A_3	
x_1	X_{10}	S_1	S_3	S_2	W
x_2,x_3	X_{20}	S_1	S_3	S_3	W
x_4,x_5	X_{30}	S_1	S_2	S_2	A
x_6	X_{40}	S_1	S_2	S_3	A
x_7	X_{50}	S_1	S_1	S_1	F_1
x_8,x_9	X_{60}	S_1	S_3	S_1	F_1
x_{10}, x_{11}	X_{70}	S_2	S_1	S_1	F_2
x_{12},x_{13},x_{14}	X_{80}	S_2	S_1	S_3	F_2
x_{15}	X_{90}	S_2	S_2	S_1	F_2

$$A_2{:}H = (3/9)(-(2/3)*\log_4(2/3)-(1/3)*\log_4(1/3))+(3/9)(-(2/3)*\log_4(2/3)$$
$$-(1/3)*\log_4(1/3)) + (3/9)(-(2/3)*\log_4(2/3)-(1/3)*\log_4(1/3))$$
$$A_2{:}H = (9/9)(-(2/3)*\log_4(2/3)-(1/3)*\log_4(1/3))$$
$$A_2{:}H = 0.459018$$
$$A_3{:}H = (2/9)(-(1/2)*\log_4(1/2)-(1/2)\log_4(1/2))-(4/9)((2/4)\log_4(2/4)$$
$$-2/4\log_4(2/4))+(3/9)*(1/3\log_4(1/3)+1/3\log_4(1/3)+1/3\log_4(1/3))$$
$$A_3{:}H = -(6/9)*(\log_4(1/2)-(1/3)\log_4(1/3)$$
$$A_3{:}H = -(2/3)\log_4(1/2)-(1/3)\log_4(1/3)$$
$$A_3{:}H = 0.597494$$

Thus we select the attribute A_2 as the first decision node since it has the minimum entropy compared to the entropy of the other attributes. The

Table E.4. Modified Table

Inputs	A_1	A_2	A_3	Outputs
x_1	S_1	S_3	S_2	W
x_2	S_1	S_3	S_3	W
x_6	S_1	S_3	S_1	F_1

Table E.5. Modified Table

Inputs	A_1	A_2	A_3	Outputs
x_5	S_1	S_1	S_1	F_1
x_7	S_2	S_1	S_1	F_2
x_8	S_2	S_1	S_3	F_2

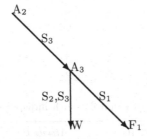

Fig. E.1. A decision tree

decision tree now has a root which is A_2. This node has three branches: S_1, S_2, and S_3. Under the branch S_3 we can find the entropy of the attributes A_1 and A_3 and choose the attribute with the minimum entropy. We repeat the process for the other 2 branches S_2 and S_1. Here are the results of the calculations: for $A_2 = S_3$ and since we have 2 possible outputs this time, the log is a log base 2 (see table E.4).

$$A_3 : H = 0$$
$$A_1 : H = -(2/3) \log_2(2/3) - 1/3 \log_2(1/3)$$
$$A_1 : H = .92$$

A_3 minimizes the entropy. Thus the shape of the decision starts to look like the one in figure E.1 is: for $A_2 = S_1$ (see table E.5) and since we have 2 possible outputs this time, the log is a log base 2.

$$A_1 : H = 0$$
$$A_3 : H = -(2/3) * ((1/2) \log_2(1/2) - 1/2 \log_2(1/2)) - (1/3) \log_2(1/1)$$
$$A_3 : H = (2/3) \log_2(1/2) = 0.6666$$

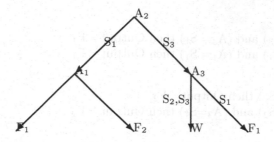

Fig. E.2. A modified decision tree

Table E.6. Modified Table

Inputs	A_1	A_2	A_3	Outputs
x_3	S_1	S_2	S_2	A
x_4	S_1	S_2	S_3	A
x_9	S_2	S_2	S_1	F_2

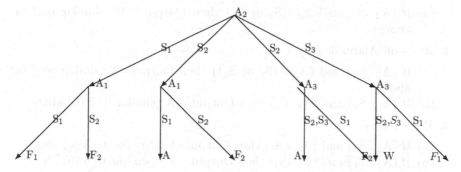

Fig. E.3. Decision Tree using Information Theory

A_1 minimizes the attribute $A_2 = S_1$. The decision tree now has the branch $A_2 = S_1$ added to it and looks like the one in figure E.2: for $A_2 = S_2$, then we have the following table (E.6):

$$A_1 : H = 0$$
$$A_3 : H = 0$$

A_1 and A_3 both minimize the entropy in this case. Thus both are possible attributes for the decision node:

Rules extracted from Figure E.3:

1. Rules on Working or W:

   (a) If $(A_2 = S_3)$ and $(A_3=(S_2$ or $S_3))$ then Output $= W$

2. Rules on Alarm or A:

   (a) If $(A_2 = S_2)$ and $(A_3= (S_2$ or $S_3))$ then Output $= A$
   (b) If $(A_2 = S_2)$ and $(A_1=S_1)$ then Output $= A$

3. Rules on F1:

    (a) If $(A_2=S_3)$ and $(A_3= S_1)$ then Output $=F_1$

    (b) If $(A_2=S_1)$ and $(A_1= S_1)$ then Output $=F_1$

4. Rules on F2:

    (a) If $(A_1 = S_2)$ then Output $=F_2$

    (b) If $(A_2 = S_2)$ and $(A_3=S_1)$ then Output $=F_2$

E.2 Application of Rough Set Theory to a Production Line Problem

The authors Gento and Redondo [[40]] applied Rough Set theory to the same example above and discovered the following rules:

Rules extracted by applying Rough Set theory to Table E.3 (see Figure E.4):

1. Rules on Working or W:

    (a) If $(A_2=S_3)$ and $A_3=(S_2$ or $S_3)$ then Output $=W$ (similar to 1.(a) above)

2. Rules on Alarm or A:

    (a) If $(A_2=S_2)$ and $(A_3= (S_2$ or $S_3))$ then Output$=A$ (similar to 2.(a) above)

    (b) If $(A_2=S_2)$ and $(A_1=S_1)$ then Output $=A$ (similar to 2.(b) above)

3. Rules on F_1:

    (a) If $(A_2=S_3)$ and $(A_3= S_1)$ then Output$=F_1$ (similar to 3.(a) above)

    (b) If $(A_2=S_1)$ and $(A_1=S_1)$ then Output $=F_1$ (similar to 3.(b) above)

    (c) If $(A_1=S_1)$ and $(A_3=S_1)$ then Output $=F_1$ (1 extra rule)

4. Rules on F_2:

    (a) If $(A_1=S_2)$ then Output$=F_2$ (similar to 4. (a) above)

    (b) If $(A_2=S_2)$ and $(A_3=S_1)$ then Output $=F_2$ (similar to 4. (b) above)

    (c) If $(A_2=S_1)$ and $(A_3=S_3)$ then Output $=F_2$ (1 extra rule)

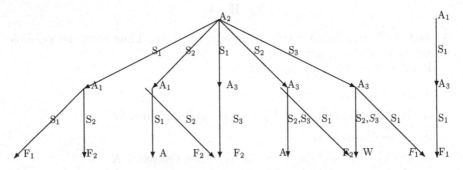

Fig. E.4. Decision Tree using using Rough Set theory

Index